Principal Facts of the Earth's Magnetism and Methods of Determining the True Meridian and the Magnetic Declination

You are holding a reproduction of an original work that is in the public domain in the United States of America, and possibly other countries. You may freely copy and distribute this work as no entity (individual or corporate) has a copyright on the body of the work. This book may contain prior copyright references, and library stamps (as most of these works were scanned from library copies). These have been scanned and retained as part of the historical artifact.

This book may have occasional imperfections such as missing or blurred pages, poor pictures, errant marks, etc. that were either part of the original artifact, or were introduced by the scanning process. We believe this work is culturally important, and despite the imperfections, have elected to bring it back into print as part of our continuing commitment to the preservation of printed works worldwide. We appreciate your understanding of the imperfections in the preservation process, and hope you enjoy this valuable book.

COAST AND GEODETIC SURVEY OFFICE.

DEPARTMENT OF COMMERCE AND LABOR
U.S. COAST AND GEODETIC SURVEY
O. H. TITTMANN, Superintendent

PRINCIPAL FACTS OF THE EARTH'S MAGNETISM

AND METHODS OF DETERMINING THE TRUE MERIDIAN AND THE MAGNETIC DECLINATION

[Reprinted from United States Magnetic Declination Tables and Isogonic Charts for 1902]

WASHINGTON
GOVERNMENT PRINTING OFFICE
1909

CONTENTS.

	Page.
PREFACE	7
DEFINITIONS	9

PRINCIPAL FACTS RELATING TO THE EARTH'S MAGNETISM.

Early History of the Compass.

Discovery of the Lodestone	11
Discovery of Polarity of Lodestone	12
Introduction of the Compass	15
Improvement of the Compass by Petrius Peregrinus	16
Improvement of the Compass by Flavio Gioja	20
Derivation of the word Compass	21
Voyages of Discovery	21
Compass Charts	21

Birth of the Science of Terrestrial Magnetism.

Discovery of the Magnetic Declination at Sea	22
Discovery of the Magnetic Declination on Land	25
Early Methods for Determining the Magnetic Declination and the Earliest Values on Land	26
Discovery of the Magnetic Inclination	30

The Earth, a Great Magnet.

Gilbert's "De Magnete"	34

The Variations of the Earth's Magnetism.

Discovery of Secular Change of Magnetic Declination	38
Characteristics of the Secular Change	40
Diurnal Variation	47
Annual Variation	52
Minor Periodic Fluctuations	53
Magnetic Storms	53
Magnetic Observatories	56

Magnetic Charts.

Isogonic Lines	62
Magnetic Meridians	63

Magnetic Surveys.

General Remarks	65
Historical Summary	67
Magnetic Survey of the United States	70

The Earth's Magnetic Poles and Magnetic Moment.

Magnetic Poles	73
Magnetic Moment	76

DETERMINATION OF THE TRUE MERIDIAN AND THE MAGNETIC DECLINATION.

Determination of the True Meridian.

By Observations on Polaris	79
By Observations on the Sun	92

Determination of the Magnetic Declination.

With an Ordinary Compass or Transit	96
With a Magnetometer	96

ILLUSTRATIONS.

FIGURES.

		Page.
1.	A Japanese South-pointing Cart (seventh century A. D.)	13
2.	Floating Compass used by Peregrinus (1269)	19
3.	Double Pivoted Compass invented by Peregrinus (1269)	19
4.	Lines of Equal Magnetic Declination for 1500 (van Bemmelen)	23
5.	Compass Sun-dial showing Earliest Magnetic Declination at Paris (1541)	25
6.	First Dip Circle (Norman's, 1576)	32
7.	Norman's Experiment showing Action of the Earth on a Magnetic Needle	33
8.	Comparison of the Secular Change Curves of the Magnetic Declination at various Stations in the Northern Hemisphere	44
9.	Curves showing Secular Change in Magnetic Declination and Dip at London, Boston, and Baltimore	45
10.	Comparison of Curve showing Change in Magnetic Declination and Dip along Parallel of Latitude 40° North in 1885, with Curve showing Secular Change at Rome	46
11.	Diagram showing Diurnal Variation of the Magnetic Declination at Baldwin, Kansas., 1901	48
12.	Coast and Geodetic Survey Magnetic Observatory at Cheltenham, Maryland	57
13.	Eschenhagen Magnetograph at Coast and Geodetic Survey Magnetic Observatory, Baldwin, Kansas	59
14.	Magnetograms showing Guatemala Earthquake Disturbance at Cheltenham Magnetic Observatory, April 18, 1902	60
15.	Magnetic Disturbance at Cheltenham Magnetic Observatory, April 10-11, 1902	61
16.	Magnetic Disturbance at Cheltenham Magnetic Observatory at time of Martinique Volcanic Eruption, May 8, 1902	61
17.	Lines of Equal Magnetic Declination for 1600 (Hansteen)	62
18.	" " " " " " 1700 (Halley)	62
19.	" " " " " " 1800 (Hansteen)	62
20.	" " " " " " 1858 (British Admiralty)	62
21.	" " " " " " 1905 (" ")	64
22.	" " " " Dip " 1905 (" ")	64
23.	Magnetic Meridians for 1836 (Duperrey)	64
24.	Lines of Equal Magnetic Declination in the Polar Regions for 1885 (Neumayer)	64
25.	Mean Secular Change of the Magnetic Declination, 1890-1900 (Neumayer)	66
26.	Map of Region around North Magnetic Pole (Schott, 1890)	75
27.	Diagram of principal Stars in the Constellations Cassiopeia and Great Bear	85
28.	Coast and Geodetic Survey Magnetometer	97

PREFACE.

The present publication is a reprint of the first part of the United States Declination Tables for 1902,[a] giving the principal facts relating to the Earth's magnetism, and the determination of the true meridian and the magnetic declination. The tables, giving the times of culminations and elongations of Polaris and its azimuth at elongation, originally published as Coast and Geodetic Survey Bulletin No. 14,[b] have been revised and extended. The tables of azimuth and apparent altitude of Polaris at any hour angle, originally published in 1896,[c] have been revised and extended to meet present requirements.

As the editions of these publications had either become exhausted or were in need of revision, their reissue was necessary to meet the continued demand for general information relating to the Earth's magnetic phenomena, and for the tables of the azimuth of Polaris which are constantly being put to practical use, especially by county engineers and surveyors throughout the country.

In connection with what is given in the text under the heading of "Discovery of the Magnetic Declination on Land" (page 25), it may now be added that since the appearance of Hellmann's paper, in 1897, A. Wolkenhauer[d] has called attention to the existence of three sun-dial compasses made previous to the year 1492. The most interesting one of these three sun dials, the one bearing date 1451, has recently been examined and described by Hellmann[e] and further confirmed his opinion that the *magnetic declination on land* was known before the first voyage of Columbus. In reviewing Hellmann's paper concerning the sun dial of 1451, Bauer[f] has recently pointed out the possibility "that this additional find, valuable as it is, is simply another proof that while the divergence of the compass direction from the true meridian may have been noted, *the divergence was not recognized at that time as a scientific fact, but as an error of the instrument.*"

---

[a] United States Magnetic Declination Tables and Isogonic Charts for 1902, and Principal Facts Relating to the Earth's Magnetism, by L. A. Bauer, Washington, 1902. Second Edition, Washington, D. C., 1903.

[b] Approximate Times of Culminations and Elongations and the Azimuths at Elongation of Polaris for the Years between 1889 and 1910, by Charles A. Schott, Bulletin No. 14, Washington, 1890.

[c] Tables of Azimuth and Apparent Altitude of Polaris at Different Hour Angles, by G. R. Putnam, App. 10, Coast and Geodetic Survey Report for 1895, Washington, 1896.

[d] Wolkenhauer, A. Beiträge zur Geschichte der Kartographie und Nautik des 15. bis 17. Jahrhunderts, Inaug. Diss. Un. Göttingen, Munich, 1904.

[e] Hellmann, G. Über die Kenntniss der magnetischen Deklination vor Christoph Columbus. Met. Zeits., April, 1906, pp. 145-149.

[f] Bauer, L. A. Earliest Values of the Magnetic Declination. Terrestrial Magnetism, Vol. XIII, pp. 97-104.

In reference to the expedition of Captain Roald Amundsen to the north magnetic pole, mentioned on page 74 of the text, it is, of course, now to be said that the expedition proceeded in 1903 as stated, and not only succeeded in reaching the supposed region of the north magnetic pole, in the vicinity of which a long series of continuous magnetic observations were carried out, but also succeeded in making the Northwest Passage, reaching the west coast of the United States by way of Bering Sea in the fall of 1906. Amundsen's location of the north magnetic pole has not yet been definitely announced, as the reduction of his observations is not completed, but from various accounts his observations will apparently place the north magnetic pole on King William Land, somewhere in the region of Ross's location, Boothia Felix.

No additional definite information in reference to the position of the south magnetic pole has been published since the preparation of the text in 1902. From the notes of the voyage of the antarctic ship "Discovery" (1901-1904),[a] it is recorded that the exploring party was due south of the south magnetic pole when in latitude about 77° 45' S. and longitude about 153° 45' E. Additional light will doubtless be shed on the question of the position of the south magnetic pole when the complete magnetic observations of this expedition are published.[b]

---

[a] The Voyage of the "Discovery," by Captain R. F. Scott, R. N., 2 vols., London, 1905, Vol. II, p 258.

[b] According to Journal of Terrestrial Magnetism, Vol. XIII, No. 4, page 186, Mr. L. C. Bernacchi recently presented a paper before the British Association for the Advancement of Science, in which he discussed some of the "Discovery" physical observations. From all the declination observations he deduced for the position of the south magnetic pole: Latitude 72° 50' S. and longitude 156° 20' E. The inclination observations indicated for the probable position of the pole: Latitude 72° 52' S. and longitude 156° 30' E., a remarkably close agreement.

DEFINITIONS.

To avoid the confusion arising from the use or misuse of the term "variation of the compass," the following terms are used instead throughout this publication:

Magnetic declination: The angle by which the compass needle points to the east or west of true north.

Secular change of the magnetic declination: The change in the magnetic declination with the lapse of years.

PRINCIPAL FACTS RELATING TO THE EARTH'S MAGNETISM.

EARLY HISTORY OF THE COMPASS.

DISCOVERY OF THE LODESTONE.

Many centuries before the Christian era writers referred to a mysterious "stone" possessing remarkable properties, chief of which was its power to "draw to it the all-conquering iron." Its earliest names appear to be Hercules stone (heraclein stone), magnet-stone, Lydian stone, siderit (iron stone), and also briefly "stone." Later the term "stone" and "Hercules stone" gave way to the name "magnet."

The precise derivation of the term "magnet," which has now become the most common one, is difficult to ascertain. Lucretius (99–55 B. C.) says it was called "magnet" from the place from which it was obtained—"in the native hills of the Magnesians." However, Pliny (23–79 A. D.) relates a prettier legend, as copied from the poet Nicander (second century B. C.), that the shepherd, "Magnes" by name, while guarding his flock on the slopes of Mount Ida, suddenly found the iron ferrule of his staff and the nails of his shoes clinging to a "stone," which became known after him as the "Magnes stone" or magnet.

The fundamental property of the lodestone of attracting iron was certainly known to the Greeks toward the close of the seventh century B. C., as it is mentioned by Thales, who lived between 640–546 B. C.

Magnetic mountains which caused ships to fall to pieces by drawing from their sides the iron nails, or, by disturbing the compass, caused to be dashed to pieces on the rocks the vessel that was unlucky enough to come within too close proximity to their influence, remained in the category of sea terrors until but a comparatively short time ago.

In writings of the middle ages we find used for the term magnet "adames," which also meant diamond; e. g., in the oriental history of the Cardinal Jaques de Vitry, of about the year 1218. The Italian term was "calamita;" the Dutch, "magnetsteen;" and "zehl-steen" (sailing stone); the Icelandic, "leider-steen" (lead stone), from which the English term of lodestone[a] is derived; the Hungarian, "magnet-ko" (magnet stone); the Polish, "magnes" and "magnet;" the Croatian, "zelezoolek" (which attracts iron); the Dalmatian, "zoosdotegh" (which draws nails); the French, "aimant" (loving stone); the Spanish, "piedramant;" and the German, "magness," "siegel-stein," and "magnetstein." The lodestone was also called by early English writers "adamant stone."

---

[a] Also spelled *loadstone*, the spelling used in this publication being the preferable one, however, as more clearly showing the derivation.

Klaproth remarks that nearly all of the European terms, as far as their signification is concerned, recur in the Asiatic tongues. Thus the most common expression of the Chinese was "thsee schy" (or loving stone), hence similar to that of the French. For example, the author Tschlin-thsangkhi says: "The magnet draws to it the iron as the tender mother calls her children to her, and for this reason it has received its name of the loving stone." Other Chinese terms for magnet were "tchu chi" (stone which deflects), "hie thy schy" (stone which unites with iron), etc.

The lodestone or natural magnet is known to the geologist as the mineral "magnetite" and is the magnetic or black oxide of iron, Fe_3O_4, this oxide being formed when iron is oxidized at a high temperature in the air, in oxygen or in aqueous vapor. It is quite widely distributed over the earth, some of the most notable specimens coming from Magnet Cove, Arkansas. Its general color is blackish or brown and occasionally grayish, and its specific gravity is 5.0 to 5.1

DISCOVERY OF POLARITY OF LODESTONE.

Not only does the lodestone or magnetite possess the property to "draw" to it iron objects, but it also has that of "polarity," i. e., it exhibits contrary effects at opposite ends, e. g., at one end it attracts the north end of a magnetic needle and at the other end repels it.

By virtue of this polarity and the fact that the "earth itself acts like a great magnet," a lodestone pointed at the ends and suspended so as to turn freely will set itself in an approximately north-and-south direction. This "directive" tendency of the lodestone or needle was termed by Gilbert in 1600 its "verticity."

It is this property of polarity which distinguishes a piece of nonmagnetized iron from a magnetized one, the former attracting either end of a compass needle, while the latter will either attract or repel, according as the unlike "poles" or the like "poles" of magnet and needle are brought together.

This property became known to European nations about the twelfth century. The Chinese are, however, generally credited with the earliest knowledge of the directive property of the lodestone and of its power to communicate polarity to iron. Tradition would even ascribe this knowledge to them as far back as the year 2634 before the Christian era. A quaint legend relates that in the sixty-fourth year of the reign of Ho-ang-ti (2634 B. C.), the Emperor, Hiuan-yuan, or Ho-ang-ti, attacked the rebel, Tchi-yeou, or Khiang, on the plains of Tchou-lou. Khiang getting the worst of the conflict, raised a great fog in order to throw the ranks of his adversary into confusion. Hiuan-yuan, however, was equal to the occasion and constructed a chariot (Tchi-nan), *which indicated the south*, so as to distinguish the four cardinal points, and thus was enabled to pursue Khiang and take him prisoner.[a]

Benjamin[b] considers this legend as "clearly mythical" and remarks that "while the beginning of Chinese history is placed by De Lacouperie at the twenty-third century B. C., other Chinese annalists regard it as impossible to rely upon any records

---

[a] Klaproth, Lettre à M. le Baron Humboldt sur l'invention de la Boussole, Paris, 1834. Also Mailla, Histoire générale de la Chine, tome I, p. 316, Paris, 1777.

[b] Intellectual Rise in Electricity, London, 1895; republished by Wiley & Sons, New York, in 1898, under the title of "History of Electricity." The writer has made considerable use of this work.

dating back more than 800 years before our era. Legge fixes the beginning of trustworthy chronology at 826 B. C., and Plath at 841 B. C. It is apparent, therefore, that in dealing with the legends and traditions which form the basis for the assertion of knowledge of the magnet by the Chinese at very ancient epochs, the doubt whether they properly belong to mythology or to history is unavoidable."

In Japan these south-pointing carts were known in the second half of the seventh century A. D. Figure 1 is a reproduction of a picture contained in Vol. XXIII of the large Japanese encyclopedia and taken from Urbanitzky's book Electricität im Alterthume, to which the writer is indebted for various references.

Several other references to the compass have been cited as appearing in early Chinese records. The first direct statement as to their knowledge of the property of *polarity* is said to have occurred in a Chinese dictionary completed about 121 A. D., a period when at least the *attractive* properties of the lodestone had been known to European nations for six centuries and more. According to Benjamin, "this statement consists of but six Chinese characters in the dictionary Choue-Wen, where the character 'Tseu' is defined as 'the name of a stone with which the needle is directed.' Even this is known only by citations in later works."

Whatever doubt may be raised regarding these early Chinese references, the fact is that the lodestone, or magnetite, is known to have existed in the iron deposits extensively worked in Shensi in 220 B. C., so that the Chinese had ample opportunities for becoming familiar with the properties of the lodestone.

The first reference to the use of the compass for navigational purposes is found in the Chinese encyclopedia, Poei-wen-yun-fou. It is said that under the Tsin dynasty, or between 265 and 419 A. D., "there were ships indicating the south."

The most remarkable passage, however, occurring in the early Chinese literature is one toward the end of the eleventh century of the Christian era in a work entitled "Mung-Khi-pi-than," viz:[a]

FIG. 1.—A Japanese south-pointing cart (7th century A. D.).

"The soothsayers rub a needle with a magnet stone, so that it may mark the south; however it declines constantly a little to the east. It does not indicate the south exactly. When this needle floats on the water it is much agitated. If the finger-nails touch the upper edge of the basin in which it floats they agitate it strongly; only it continues to slide, and falls easily. It is better in order to show its virtues in the best way to suspend it as follows: Take a single filament from a piece of new cotton and attach it exactly to the middle of the needle by a bit of wax as large as a mustard seed. Hang it up in a place where there is no wind. Then the needle constantly shows the south; but among such needles there are some which, being rubbed, indicate the north. Our soothsayers have some which show south and some which show north. Of this

---

[a] Ed. Biot: Comptes rendus, t. XIX, p. 825. The passage is quoted from Benjamin's book.

property of the magnet to indicate the south, like that of the cypress to show the west, no one can tell the origin."

According to Klaproth, the same fact is related in a natural history compiled by Kew-tsung-schy, in the years 1111–1117, under the title of Pen-thsao-yan-i, and it is stated that the "south end of the needle always shows a deviation toward the point 'ping,' *i. e.*, E. ⅙ S.," hence ⅙ of 90° or 15° east of south, so that the north end pointed 15° west of north.

Benjamin says "that the tendency of the magnetic needle to depart from the true north appears to have been observed by the Chinese geomancers in the compasses used by them long before any marine use of the instrument was made. A so-called life of Yi-hing, a Buddhist priest and imperial astronomer, undertakes to show that the 'variation' in the eighth century was nearly 3° west of south. Later we find the geomancers adding special circles of symbols to the compass card, such as a circle of nine fictitious stars, a circle of sixty dragons, and so on, and, among these, circles of points especially constructed to allow for 'variation'. This was done in the year 900 by Yang Yi when the variation was 5° 15' east of south, and again three centuries later when it had increased to 7° 30'[a] in the same direction."

The Chinese apparently would have to be credited by these passages with a knowledge of the properties of the magnet far superior to that possessed at that period by the European nations. They seem not only to have known of the magnetic declination of the needle, by reason of which the needle did not point true north and south, but also to have anticipated Europeans by several centuries in the most delicate method of suspension of a needle by means of a fiber. The Jesuit Lana, according to Hansteen, is said to have introduced the fiber suspension in Europe about 1686. According to Prof. Sylvanus P. Thompson, however, the suspending of a magnetic needle by a thread occurs in the Speculum Lapidum of Camillus Leonardus, published at Venice in 1502.

Klaproth, who made a special study of the early history of the compass, found "no indubitable use" of the compass by the Chinese in *navigation* until toward the end of the thirteenth century, at which time it had been on European ships for a century or more. All efforts to satisfactorily account for the spread of the knowledge of the properties of the lodestone from Eastern to Western nations, or vice versa, have thus far failed.[b]

Summing up all the evidence, it would seem that the prime properties of the lodestone—attraction, polarity, directivity—were doubtless discovered independently by the Chinese and by the occidental peoples and that the preponderance of evidence of priority at present would seem to be on the Chinese side.

The Chinese undoubtedly were the first to use the compass in land journeys and in the orientation of buildings and sites. It is related that, in the early part of the four-

---

[a] According to Klaproth, as cited above, this was 15°.

[b] The number of points of the compass, according to the Chinese, is twenty-four, which are reckoned from the south pole; the form also of the instrument they employ is different from that familiar to Europeans. The needle is peculiarly poised, with its point of suspension a little below its center of gravity, and is exceedingly sensitive; it is seldom more than an inch in length and is less than a line in thickness. It appears thus sufficiently evident that the Chinese are not indebted to Western nations for their knowledge of the use of the compass. (Encyclopædia Britannica, 9th ed., art. Compass.)

teenth century (1314–1320 A. D.), the temple of Yao-mu-ngan was oriented in this way. It is an interesting fact that they were guided by the south end of the needle, their name for compass being "ting-nan-ching," or needle pointing to the south. This was probably because in China the south is considered the honorable quarter, the Emperor taking his position facing south, and prominent buildings being placed facing south. To distinguish the south end of the needle from the north end it was painted red.

INTRODUCTION OF THE COMPASS.

The earliest definite mention at present known of the use of the compass in the Middle Ages occurs in a treatise entitled "De Utensilibus," written toward the end of the twelfth century by an English monk, Alexander Neckam. He says:

"The sailors, moreover, as they sail over the sea, when in cloudy weather they can not longer profit by the light of the sun, or when the world is wrapped in the darkness of the shades of night, and they are ignorant to what part of the horizon the prow is directed, place the needle over the magnet, which is whirled round in a circle, until, when the motion ceases, the point of it (the needle) looks to the north."

Soon after the introduction of the compass, laws were framed against the falsifying of the compass. One of the most common beliefs which prevailed for many centuries was known as the "garlic myth," and mariners were charged not to eat onions or garlic lest the odor "deprive the stone of its virtue by weakening it and prevent them from perceiving their correct course."[a]

In a poem entitled "Love's Complaint," found by M. Paulin Paris, a distinguished antiquarian, in a MS. of 1329 which he attributed to William the Clerk, a vassal of Sire Rauf or Raul, who fought in the wars of Frederick I in Italy (1159 to 1177) appears the following description of the compass used at that time:

> Who would of his course be sure,
> When the clouds the sky obscure,
> He an iron needle must
> In the cork wood firmly thrust.
> Let the iron virtue lack
> Rub it with the lodestone black
> In a cup with flowing brim,
> Let the cork on water swim,
> When at length the tremor ends,
> Note the way the needle tends:
> Though its place no eye can see—
> There the polar star will be.

Furthermore, in the preceding verse he appears to assign the cause for the north and south pointing of the needle to the attractive influence of the polar star, a belief current until Gilbert's time (1600).

---

[a] One of these laws was as follows: "Whoever, being moved by sedition, shall menace the master or pilot of a ship with the sword, or shall presume to interfere with the nautical gnomon or compass, and especially, shall falsify the part of the lodestone upon which the guidance of all may depend, or shall commit like abominable crimes in the ship or elsewhere, shall, if his life be spared, be punished by having the hand which he most uses fastened, by a dagger or knife thrust through it, to the mast or principal timber of the ship, to be withdrawn only by tearing it free." (Benjamin's Intellectual Rise in Electricity.)

Allusions to the compass among the early writers now began to multiply, e. g., Guyot de Provins, in a poem written 1203-1208, Cardinal de Vitry (1218), and others. In a poem by Guido Guinicelli, an Italian priest who died in 1276, the following suggestive lines occur:

> In what strange regions 'neath the polar star
> May the great hills of massy lodestone rise,
> Virtue imparting to the ambient air
> To draw the stubborn iron; while afar
> From that same stone the hidden virtue flies
> To turn the quivering needle to the Bear,
> In splendor blazing in the Northern skies.

Matthew Paris, in relating the sending of the first papal legate to Scotland in 1247, says he "drew the money out of the Scots to himself as strongly as the adamant does iron."

In the middle of the thirteenth century the compass was in regular use among the Norwegians.

Bacon appears to have had his attention directed to the lodestone, which he calls "the miracle of nature," by Glanvil's encyclopedic work, written about 1250. He says "that the iron which is touched by the lodestone follows the part of the latter which excites it, and flies up from the other part, and that it turns to the part of the heavens to which the part of the magnet wherewith it was rubbed conforms." Furthermore, "that it is not the polar star which influences the magnet, for if such were the case the iron would always be attracted toward the star; on the contrary, the rubbed portion of the iron will follow the rubbed part of the magnet in any direction, backward or forward, or to the right or left," etc.[a]

IMPROVEMENT OF THE COMPASS BY PETRIUS PEREGRINUS.

We now turn to one of the most famous of the writings of the Middle Ages. Bacon in his "Opus tertium" says "there are but two perfect mathematicians, Master John of London and Master Petrius de Maharn, curia, a Picard." Peter stands especially high in his estimation. He was the author of the famous letter known as "Epistola Petri Peregrini de Maricourt ad Sygerum de Foucaucourt militem de Magnete."

This letter "on the magnet," written by the nobleman Pierre de Maricourt on August 12, 1269, to his friend and neighbor Syger de Foucaucourt, is probably the oldest European treatise on the magnet. The author's surname "de Maricourt" is derived from a little village in Picardy, France, from whence he came. He is, however, more generally referred to as "Petrius Peregrinus," the appellation of Peregrinus or Pilgrim indicating that he had taken part in the Crusades. He was a partisan of Charles of Anjou, who had been crowned King of the two Sicilies by Pope Urban IV, and who was laying siege for a second time to the town of Lucera, situated in the province of Apulia in southern Italy. Under the walls of this town Peregrinus wrote his memorable "epistola," which became known to many of the learned men of his time and succeeding centuries and had considerable influence on early writers on magnetism. It was reproduced in 1558 with an introduction and comments by Achilles Gasser, a

---

[a] If this quotation be correct as taken from Benjamin, then the latter part of Bacon's statement, "that the iron turns to the part of the heavens to which the part of the magnet wherewith it was rubbed conforms," is incorrect. The contrary, as we shall see later, is the case.

physician of Lindau, Germany, and again by subsequent authors, and more recently by Hellmann in his excellent series of Berlin reprints, "Neudrucke"—Rara Magnetica No. 10.[a]

Peregrinus was a man of learning, had the academic title of "magister," and, as stated, was regarded very highly by his contemporary, Roger Bacon. The deductions in his letter reveal in general a clear insight and sound reasoning powers. They are based usually on actual experiment, which doubtless accounts for the influence his little treatise exerted.

Some of the facts which Peregrinus cites in his letter had been previously known. However, he appears to have had the faculty of putting them in precise language. A summary of the contents of the letter will be found in Benjamin's book, from which the quotations below have been taken.

Peregrinus, in direct contradiction to the earlier writers, who invariably preferred the lodestone from India, gives preference to the one from northern Europe, which was used principally by sailors in the northern seas.

He explains how the poles of a magnet may be found, thus:

"The stone is to be made in globular form and polished in the same way as are crystals and other stones. Thus it is caused to conform in shape to the celestial sphere. Now place upon it a needle or elongated piece of iron, and draw a line in the direction of the length of the needle, dividing the stone in two. Then put the needle in another place on the stone, and draw another line in the same way. This may be repeated with the needle in other positions. All of the lines thus drawn will run together in two points, just as all the meridian circles of the world run together in two opposite poles of the world."

Peregrinus probably first found the poles in the way that is above described. Then, afterwards, he remarked that at the points so determined the needle was more strongly attracted than elsewhere. Consequently, he sees that the poles can be detected without marking the meridians by simply noting the places on the stones where the needle is most frequently and powerfully drawn. "If, however," he continues, "you wish to be precise, break the needle so as to get a short piece about two nails in length. Place this on the supposed polar point. If the needle stands perpendicularly to the surface of the stone such point is the true pole; if not, then move the needle about until the place is found where it does thus stand erect. If these points are accurately ascertained and the stone is homogeneous and well chosen," he adds, "they will be drawn diametrically opposite one another, like the poles of the sphere."

If the Earth's magnetism were uniformly distributed, Peregrinus's method of "converging magnetic meridians" could be applied to determine with greater accuracy, and certainly with more comfort, the position of the Earth's magnetic poles than by specially equipped expeditions to the arctic and antarctic regions. It would suffice to select a few well-chosen stations in easily accessible and climatically comfortable regions, to determine accurately the magnetic declination of the needle at these points, and to determine by an easy computation the points of intersection of the great circles passing through the compass directions at the selected stations. It will be seen,

---

[a] Sparing as Gilbert is in conceding the excellence of any work on magnetism prior to his own, the "De Magnete" of 1600, he characterizes Peregrinus's work "as a pretty erudite book, considering the time."

however, that owing to the great irregularity in the distribution of the Earth's magnetism this method is not admissible, and would give positions for the magnetic poles differing considerably from the actual positions.

Peregrinus next explains how to designate the two poles and to distinguish them from each other.

"'Take,'' says Peregrinus, ''a wooden vessel, round, like a dish or platter, and put the stone in it so that the two points of the stone may be equidistant from the edge; then put this in a larger vessel containing water, so that the stone may float like a sailor in a boat. The stone so placed will turn in its little vessel until the north pole of the stone will stand in the direction of the north pole of the heavens, and the south pole in that of the south pole of the heavens; and if it be removed from this position, it will return thereto by the will of God. Since the north and south parts of the heavens are known, so will they be known in the stone, because each part of the stone will turn itself to its corresponding part of the heavens.''

Then, ''If the north part of the stone, which you hold, be brought to the south part of the stone floating in the vessel, the floating stone will follow the stone you hold, as if wishing to adhere to it; and, if the south part of the held stone be brought to the north part of the floating stone, the same thing will happen. Know it, therefore, as a law,'' he says, ''that *the north part of one stone attracts the south part of another stone, and the south the north.*''

We thus have a recognition of the well-known fact that unlike magnetic poles attract each other and while Peregrinus does not explicitly state the additional fact that like poles repel each other, it stands to reason that in the course of his experiments the fact of repulsion of like poles must likewise have manifested itself to him, especially, as it was known to his master, Bacon. However, it was customary for the early writers to speak simply of the ''*attractive* virtue of the magnet.''

It will be noticed that Peregrinus designated that part of the lodestone which points to the north as the north end or pole, and that part which is directed to the south, the south pole. He says, ''You will infer what part of the iron is attracted to each part of the heavens from knowing that the part of the iron which has touched the southern part of the magnet is turned to the northern part of the sky. The contrary will happen with respect to that end of the iron which has touched the north part of the stone, namely, it will direct itself towards the south.''

This is the first clear and accurate statement regarding the character of the poles induced in the ''iron'' by its ''touch'' with the ''magnet'' or lodestone, and the quarter of the heavens to which each pole will point, if the iron be freely suspended.[a] It will be noted that Bacon's statement (p. 16) is just the reverse of that of Peregrinus.

According to the laws of magnetism, the part of the iron touched by the magnet or lodestone will have induced in it a magnetic pole of an opposite kind to that in the end of the magnet used. Furthermore, since like poles repel and unlike ones attract each other, it is manifest that if the north end of a compass is called the *north* pole, the magnetism in the northern regions of the earth must be of the *south* pole kind, otherwise we should have repulsion instead of attraction. Or, if in the north end of the compass there resides magnetism of the *south* pole kind, then the earth's north magnetic

---

[a] Benjamin, thinking that Peregrinus had committed an error in his statement, offers various apologies for him.

pole has magnetism of the *north* pole kind. To avoid this confusion the north end of the compass is frequently referred to as "the north-seeking or north-pointing end," and the south end as the "south-seeking or south-pointing end." The part of the "iron," then, which touches the north-seeking end of the magnet will have magnetism of the south-seeking kind induced in it, and will point or be attracted to the south when the iron is delicately supported, and the part which is rubbed by the south-seeking end of the magnet has induced in it a pole of the north-seeking kind and hence will point to the north.

The chief achievement of Peregrinus was his improvement of the mariner's compass, which at that time was a very crude contrivance indeed, the magnet being supported by a reed floating in a vessel of water, and provided neither with an index to reckon from nor with a compass card. He combined the compass with the nautical astrolabe for measuring the sun's altitude, provided a fiducial line, or the so-called "lubber's point," and a graduated scale, thus enabling the mariner not only to steer his ship more truly, but likewise to determine the azimuth of a heavenly body. At first he floated his compass, but later introduced for the first time the pivoted or, rather, socketed compass, the description of which, as given by Benjamin, is as follows:

"The floating bowl and the large vessel of water are abolished, and in place of them there is the ordinary circular com-

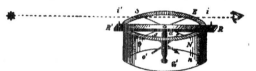

FIG. 2.—Floating compass used by Peregrinus (1269).   FIG. 3.—Double-pivoted compass invented by Peregrinus (1269).

pass box of to-day. Its edges are marked as those of the bowl were—with the degree of the circle. It is covered with a plate of glass. In the center of the instrument, and stepped in the glass cover and in the bottom of the box, is a pivot, through which passes the compass needle, now no longer an ovoid lodestone, but a true needle of steel or iron. Then at right angles to this needle is another needle, which, curiously enough, he says is to be made of silver or copper. Pivoted above the glass cover is an azimuth bar, as before, with sight pins at the ends. Now, he says, you are to magnetize the needle by means of the lodestone in the usual way, so that it will point north and south, and then the azimuth bar is to be turned on its center so as to be directed toward the sun or heavenly bodies, and in this way, of course, the azimuth is easily measured. In fact, the device is the azimuth compass of the present time. 'By means of this instrument,' says Peregrinus, 'you may direct your course toward cities and islands and all other parts of the world, either on land or at sea, provided you are acquainted with the longitudes and latitudes of those places.'"

Figure 2 represents the floating compass used by Peregrinus, and figure 3 his double-pivoted compass. Both figures have been directly reproduced from the memoirs on Peregrinus by Bertelli, who made the subject a special study.

It will be noticed that Peregrinus had in this compass all the devices necessary for ascertaining whether the magnetic needle pointed precisely to the north, or declined away from the north; however, he does not seem to have noticed any such departure. He would be especially interested in this, as he supposed that "from the poles of the world the poles of the magnet received their virtue." That he did not remark any declination indicates pretty strongly that the needle did not, at that time, point very far from north, so that if he did observe any departure, the smallness of the amount doubtless led him to ascribe it to imperfection of construction of his compass. A similar conclusion has been reached by the writer from other researches. At present the needle points about 9° west in southern Italy.

Peregrinus was credited by Thévenot in 1681 with having found a magnetic declination of 5° east in 1269, but Wenckebach's researches showed that this was an insertion in the Leyden manuscript of his "epistola," made in the early part of the sixteenth century, about which time the needle did actually point that amount at Rome. (See Table I.) Thévenot had likewise erroneously ascribed the authorship of this famous letter to "Peterus Adsigerus."

Recapitulating, Peregrinus may be accredited with very notable discoveries and achievements, chief of which are:

1. The mode of locating and distinguishing the magnetic poles of a magnet.
2. The method of touch and rubbing for reversing the polarity of a magnet and the fact that a magnet can be broken into any number of fragmentary pieces, each of which will be a magnet.
3. The first attempt at an azimuth compass, and the introduction of a mode of pivot suspension of the needle.

Improvement of the Compass by Flavio Gioja.

The contents of Peregrinus's letter did not become widely known, the few manuscript copies which had been made by the early monks lying buried in the monasteries until the sixteenth century, and so it happened that many of his discoveries were rediscovered.

In Peregrinus's pivoted compass the needle passed through a vertical shaft pivoted in the top and bottom of the compass, so that the shaft and needle turned together. In the modern compass, as is known, the compass needle turns on a fixed point. Furthermore, his compass lacked the modern subdivision of the circle into thirty-two points or the so-called "Rose of the Winds."

Flavio Gioja, who came from Positano in the hills back of Amalfi, Italy, is credited with the invention of the mariner's compass some time prior to 1318 (about 1302). Thus, Anthony of Bologna, in the latter part of the fourteenth century, writes that "Amalfi first gave to seamen the use of the magnet." It is considered probable that Gioja introduced the compass card of thirty-two points, or "Rose of the Winds," the mode of pivot suspension whereby the needle turns on a fixed point, and the attaching of the card to the compass needle, thus adding greatly to the usefulness of the compass at sea. The earliest maps having the "Rose of the Winds" are Genoese, of about the year 1318. During the summer of 1901 the invention of the mariner's compass by Gioja was celebrated at Amalfi by the Italians.

The character of the compass used in Mediterranean waters in the fourteenth century is seen from a statement of Da Buti's in 1380: "The navigators have a compass, in the middle of which is pivoted a wheel of light paper which turns on its pivot, and that on this wheel the needle is fixed and the star (Rose of the Winds) painted." The adoption of this compass by the English did not apparently take place for some time, as Chaucer does not mention until 1391 the division of the compass circle into 32 points instead of 24 points.

DERIVATION OF THE WORD "COMPASS."

The following quotation is from Prof. J. A. Fleming's lecture on "The Earth a Great Magnet," delivered at Bristol, England, in 1898:

"The word *compass* is an old English word, signifying a circle. 'My green bed embroidered with a compass' is mentioned in the will of Edward, Duke of York, who died in 1415.

"The well-known instrument for describing a circle is called a compass or pair of compasses. To encompass means to surround as by a circle, and most of you at some time or another have seen a public house with the sign of the 'Goat and the compasses,' which antiquarians tell us is only a corruption of the old pious house motto, 'God encompasses us.' Hence the magnetic instrument takes its familiar name from the circle of degrees or points which Peregrinus or Gioja added to enable it to indicate the angular distance of an object from the meridian."

VOYAGES OF DISCOVERY.

Under the initiative of Prince Henry of Portugal—Henry the Navigator—who founded a naval college, corrected charts, improved compasses, and made other advances in navigation, the compass played an important part in the great voyages of discovery of the fifteenth century. No important discovery relating to the compass resulted, however, until the memorable voyage of Columbus in 1492. Before passing to this mention should be made of the former compass charts toward the close of the fourteenth century and the first half of the fifteenth.

COMPASS CHARTS.

The early charts of the Mediterranean coasts of the fourteenth and fifteenth centuries were oriented by the compass and all bearings from one port to another were compass directions; hence these charts are known as "compass charts." It will be recalled that at their date the magnetic declination of the compass had not become known; it was believed that the compass pointed "true to the north pole," and that, hence, compass directions were also true directions. If a compass showed any marked departure from the true north this was accredited to mechanical imperfection in its construction.

The earliest of these charts were by Marino Sanuto, between 1306 and 1324. The best, however, are those in the atlas of Andrea Bianco of the year 1436 and this atlas has been subjected to a critical examination by Oscar Peschel.[a] He found that in spite

---

[a] Der Atlas des Andrea Bianco vom Jahre 1436, in zehn Tafeln. Photographische Facsimile in der Grösse des Originals, vollständig herausgegeben von Max Münster und mit einem Vorworte versehen von *Oscar Peschel*. Venedig, H. F. M. Münster, 1869.

of the crude appliances in use at that date the distances from place to place harmonized with later, more accurate determinations in a most remarkable manner, but the places were not always in their proper parallels of latitude, their departure therefrom varying in a perfectly systematic manner. Thus two places on the west Mediterranean coast were in the same parallel of latitude as places on the east Mediterranean coast, which as a matter of fact are situated in lower latitudes. In other words, the places had been plotted according to magnetic meridians and parallels. By measuring the angle for Rome, through which the chart[a] had to be turned in an ENW. direction, in order that the various places would fall in their proper *geographic* parallels, the writer found that *the magnetic declination at Rome was about 5° East in 1436 (or more likely before, since the charts were undoubtedly constructed from data obtained during many years prior to date of publication, 1436). This is the earliest information at present obtainable regarding the amount of the magnetic declination in Europe.*

BIRTH OF THE SCIENCE OF TERRESTRIAL MAGNETISM.

DISCOVERY OF THE MAGNETIC DECLINATION AT SEA.

That the needle pointed "true to the pole" of the heavens or to the pole star had been, as we have seen, the general belief up to the close of the fifteenth century. It remained for the terrorized sailors on Columbus's first voyage to the New World to be made aware of the next great fact of the Earth's magnetism, viz, that the needle changes its direction from place to place and points exactly north and south over but a very limited region of the Earth.

It will be recalled that after leaving Palos Columbus set sail for Gomera, one of the Canary Islands, whence he laid his course due west. Not many days out from Gomera, on September 13, 1492, to the great consternation of the sailors, it was noticed that "at the first of the evening of this day the needles varied to the NW., and the next morning about as much in the same direction. * * * September 17 the pilot took the sun's amplitude and found that the needle varied to the NW. a whole point of the compass. The seamen were terrified and dismayed, without saying why. The admiral discovered the cause and directed them to take the amplitude again next morning, when they found that the needles were true. The cause was that the star moved from its place, while the needles remained stationary."[b]

Before this time, as will be seen from Fig. 4, which gives the lines of equal magnetic declination for 1500, as recently drawn by van Bemmelen, the compass had pointed a few degrees east of north, but the amount, about 3° at Palos and at Gomera, was too small to attract special attention, and if it had it would have been attributed to an imperfection in the construction of the compass. The compasses used were doubtless divided into points ($11\frac{1}{4}°$) and half points, allowing quarter points (about 3°) to be estimated. (In Fig. 4 the minus sign means east declination.)

After leaving Gomera the easterly declination of the compass, it will be seen, steadily diminished, until about September 13, when it was observed in the evening to

---

[a] Bianco's chart in E. Mayer's "Die Entwickelung der Seekarten, Wien, 1877" was used.

[b] Personal Narrative of the First Voyage of Columbus to America, translated by Samuel Kettell. Published by Thomas B. Wait & Son, Boston; G. C. Carvill, New York, and Carey & Lea, Philadelphia, 1827.

pass from east to west. According to Schott's computation,[a] the flagship of Columbus was at noon on September 13, 1492, in north latitude 28° 21', and in longitude 29° 16' west of Greenwich. This position is probably not far from the place through which the line of no magnetic declination—the so-called agonic line—along which the needle did stand "true to the pole," passed at that date. This line, as is seen from Fig. 4, lay a little to the west of Fayal Island of the Azores.

It will be noticed from the above extracts that on September 17 Columbus had gone far enough west of this line to have had the compass bear a whole point ($11\frac{1}{4}°$) to the west. That the next morning "the needles were true again" is inexplicable, except that in order to allay the fears of his sailors he practiced some pardonable deception on them, and may possibly have changed the points of the compass, as he had done, according to his own confession, once before on another voyage, in order to force the inclination of a possibly mutinous crew to his will

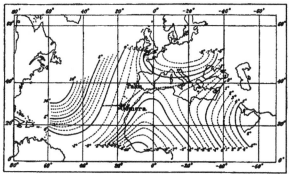

FIG. 4.—Lines of equal magnetic declination for 1500 (van Bemmelen).

The explanation which Columbus gave for the departure of the needles observed between September 13 and 17, that the North Star moved in its place, while the needles remained stationary, was, of course, a mere fiction to quiet the apprehensions of his crew. Columbus, according to the history written by his son, believed, as did Peregrinus and Bacon, that the needle was attracted or directed not by the Pole Star, but by all points of the heavens.

According to Schott's investigations, it would seem that toward the end of September, when about in midocean, the needle had reached its maximum westerly pointing; thereafter it continued to diminish, until at the first landing place of Columbus, which, according to the researches of Lieut. J. B. Murdock,[b] of the United States Navy, appears unquestionably to have occurred at Watlings Island, the needle bore but a trifle west.

---

[a] See Appendix No. 19, United States Coast and Geodetic Survey Report for 1880, p. 5, and Appendix No. 7, report for 1888, p. 305.

[b] "The Cruise of Columbus in the Bahamas, 1492." Proceedings of the U. S. Naval Institute No. 30, Annapolis, April, 1884.

Columbus himself does not mention the declination of the compass after September 17, nor does he say anything about the behavior of the compass on his return voyage, nor does he record anything regarding the compass on his second voyage (1493–1496), nor on the fourth (1502–1504). However, on the third voyage (from 1498 to 1500), he writes as follows:

"I remarked that from north to south in traversing these hundred leagues (300 geographical miles) from the said islands (Azores) the needle of the compass, which hitherto had turned toward the NE., turned a full quarter of the wind to the NW., and this took place from the time we reached that line."[a]

Continuing, he says, "For in sailing thence (from the Azores) westward the ship went on rising smoothly toward the sky and then the weather was felt to be milder, on account of which mildness the needle shifted one point of the compass; the further we went the more the needle went to the NW., this elevation producing the variation of the circle which the North Star describes with its satellites."[b]

A second point in the line of no magnetic declination, situated farther north than the one of Columbus, was found by Sebastian Cabot and dates from 1497 or 1498. He found, when on the meridian 110 miles west of the island of Flores, one of the Azores, and in latitude approximately 46° or 47°, that he was in a position where the needle had "no variation."[c]

This line along which the needle pointed exactly to the north, one point of which had been discovered by Columbus and another by Sebastian Cabot, was believed to be a convenient line, "given by nature herself," from which to reckon longitude, especially as it almost passed through the place from which longitude was then reckoned, and it figured prominently for many years in political geography as the line of demarcation between the rival kingdoms of Portugal and Castile. It can be seen, however, by referring to Fig. 4, that this line does not coincide with a true meridian and that it is

---

[a] Select letters of Columbus, 2d edition, translated and edited by H. Major, London, 1870; printed for the Hakluyt Society, pp. 131, 135.

[b] Regarding this passage Schott (App. 19, U.S. Coast and Geodetic Survey Report for 1880, p. 414) says: "It is evident that the extract from the third voyage is but an amplification of his first account, and expresses his conviction that west of the Azores, where the declination was a little easterly, it changed to the westward, being nearly zero at Corvo, and gradually increasing to one point or 11° W. at a distance of 300 nautical miles west of the longitude of Corvo. The position of Rosario, on the southeast part of the island of Corvo, is, according to the Carta Esferica de las Islas Azores, Madrid, 1855, in latitude 39° 41′ and longitude 24° 53′ west of San Fernando, or in 31° 07′ west of Greenwich (according to the Conn. des Temps); 100 leagues or 300 nautical miles west of this longitude would correspond (in latitude 28°) to 5° 40′ and would bring the Columbus line in longitude 36° 47′ W."

[c] In App. 7, U. S. Coast and Geodetic Survey Report for 1888, p. 305, second footnote, Schott says: "Soon after the discovery by Columbus of a point of no variation in the Atlantic, Sebastian Cabot discovered a second one farther north and evidently belonging to the same agonic curve. Livio Sanuto states in his Geographica Distincta (Venice, 1588) that he procured the information from Sebastian Cabot and made use of his map (probably that composed in 1544), on which the position of the meridian intersecting the point of no variation was seen to be 110 miles to the west of the island of Flores, one of the Azores; see Narrative and Critical History of America, by Justin Winsor, Vol. III, Boston and New York, 1884, p. 41. This discovery was probably made on the second voyage of the Cabots, in 1498, although it may have been noted in the first, 1497, by the elder Cabot. The latitude of the point is uncertain, but may be approximated from the fact that in the first voyage land was apparently sighted at Cape Breton, and in the second the coast of New Foundland (Baccalaos), which is said to have been made from the north."

moreover a very devious and variable line, ever changing its course and form with the lapse of time.

Thus by the end of the fifteenth century the two new facts that the compass needle does not, in general, point true north or south, but a certain amount east or west, and that the amount varies with locality, had become known among western nations; Columbus must be credited with their discovery.[a] *The necessity for measuring the angle of pointing of the needle thus became apparent in 1492, and hence this must be regarded as the year of birth of the science of terrestrial magnetism, which has for its special object the measurement of the earth's magnetic elements.*

DISCOVERY OF THE MAGNETIC DECLINATION ON LAND.

According to Hellmann,[b] "It was the construction of sundials that first brought those on land to a true perception of the declination of the magnetic needle from the

FIG. 5.—Compass sun-dial showing earliest magnetic declination at Paris (1541).

astronomical meridian" and "not the discovery of Columbus, of which nothing appeared in print." In the early part of the sixteenth century the quaint old German town of Nuremburg was quite a center for the manufacture of sundials provided with magnetic needles, which found a ready market not only in Germany but in many other countries and were widely used.

One of the most famous of these "compass makers," as the makers of these compass sundials were called, was Georg Hartmann, who lived in Nuremburg from the

---

[a] Columbus is generally credited merely with the discovery of the second fact, viz, the change of the magnetic declination from place to place. However, no satisfactory evidence has thus far come to light, as has been shown, that the first fact was known before his time, except apparently among the Chinese.

[b] "The Beginnings of Magnetic Observations," by G. Hellmann, Journal Terrestrial Magnetism, Vol. IV, pp. 73–86.

year 1518 until his death, serving as vicar of the famous church of St. Sebaldus. Hartmann lived in Rome about 1510 and appears to have made there the first observation of the magnetic declination on land, finding that the magnetic needle pointed at Rome 6° east of north. Apparently he did not make known this discovery until in a letter written March 4, 1544, to Count Albert of Prussia. In his letter he also says that at Nuremburg the needle points 10° and "at other places more or less."

Fig. 5 is a reproduction of an ivory sundial found by Le Monnier[a] in the collection of Prince de Conti and constructed by Hieronymus Bellarmartus. It shows that the needle at Paris pointed in 1541 about 7° east—this being the oldest known value at Paris.

EARLY METHODS FOR DETERMINING THE MAGNETIC DECLINATION AND THE EARLIEST VALUES ON LAND.

The earliest method was that used by Columbus of noting the magnetic bearing of the Pole Star. A Sevillian apothecary, Felipe Guillen, devised an instrument which he presented to the King of Portugal, João III, in 1525, and which he termed "*brújula de variación.*" By means of this instrument the declination was determined by noting with the aid of the shadow thrown by a stylus, the magnetic bearing of the Sun at equal altitudes before and after noon; the half difference of the bearings was the declination.

The first one who published useful methods for determining the magnetic declination appears to have been *Francisco Falero*[b] in 1535. In Hellmann's "Rara Magnetica" is reproduced the special chapter on this subject entitled "Del Nordestear de la Agujas." According to Hellmann, *in Falero's book is found the first reference in print to the magnetic declination.*

He gives the following three methods for its determination: (1) Magnetic bearing of Sun at apparent noon when the shadow of the stylus falls to the north: (2) Guillen's method, and (3) magnetic bearing of Sun at sunrise and sunset.

In 1537 Pedro Nunes improved Guillen's instrument, adding the means for measuring the Sun's altitude and inventing a new method for the determination of latitude at any time of day.

The first fairly extensive series of carefully made declinations at sea is due to João de Castro, who in 1538 commanded one of eleven ships sent to the East Indies by the Infanta Dom Luiz and who later became the fourth vice king of India. He diligently made magnetic, meteorological, and hydrographic observations on the entire voyage.[c]

The first treatise published on the subject in England was that of W. Borough: "A Discours of the Variation," London, 1581, annexed to Norman's "Newe Attractive," and republished with it three times (1585, 1596, and 1614). The methods in principle are Falero's. Borough gives in this book his observations for determining the magnetic declination at London (Limehouse), on October 16, 1580, being

---

[a] *Le Monnier:* "Histoire de l'Académie Royale de Sciences," Année, 1771, p. 29. The cut is reproduced from Hellmann's article cited above.

[b] Tratado del Esphera y del arte del marear, Sevilla, 1535.

[c] The most recent collection and utilization of the values will be found in van Bemmelen's "Die Abweichung der Magnetnadel," Batavia, 1899.

doubtless the first observations printed in detail. He deduced from these a value of 11° 15' E.[a]

The first collection of values (42) of the magnetic declination of the sixteenth century, which, however, was far from being complete, was contained in Simon Stevin's "De Havenvinding," published in Dutch, at Leyden, in 1599.[b] This was translated into Latin by Hugo de Groot (Grotius) under the title of "$\Delta\iota\mu\varepsilon\nu\tau\rho\varepsilon\tau\iota\kappa\eta$ sive portuum investigandorum ratio," and likewise published in 1599. It was translated by Edward Wright into English and published separately, and also appended to the third edition of his Errors in Navigation; the table of declinations had appeared already in the second edition of his work. The following definition of the magnetic declination taken from Grotius's translation is of interest: "Declinatio magneticæ à Septentrione ad Orientem, $\dot{\alpha}\nu\alpha\tau\lambda\iota\sigma\mu\acute{o}s$ vocatur, Occidentem versus $\delta\upsilon\sigma\iota\sigma\mu\acute{o}s$, et nomine universali $\chi\alpha\lambda\iota\beta\acute{o}\kappa\lambda\iota\sigma\iota s : \chi\alpha\lambda\upsilon\beta\acute{o}\kappa\lambda\iota\sigma\iota s$[c] ante et $\dot{o}\rho\theta o\beta o\rho\varepsilon o\delta\varepsilon\iota\xi\iota s$ generali $\chi\alpha\lambda\upsilon\beta o\delta\varepsilon i\xi\varepsilon\omega s$ nomine appellantur."

It will be seen that he used the term "magnetic declination" to denote what Norman, Borough, and, later, Gilbert termed as "variation of the compass."[d] The same writers used the word "declination" to denote what is now known as dip or "inclination." Because of this confusion of terms, careful scrutiny of the early references regarding "declination" is necessary. Instead of Grotius's terms, "anatolismos" for east declination and "dusismos" for west declination, the Dutch original has "Ostering" and "Westering," respectively, whereas Wright uses "variation west" and "variation east." The terminology of Grotius was extensively used by the seventeenth century authors of works on magnetism in the Latin language. Stevin's interesting little work owed its origin to the patronage of Count Moritz of Nassau, admiral of the Dutch fleet, who saw the great importance in navigation of accurate knowledge of the magnetic declination.

Table I represents an attempt to collect the values of the magnetic declination up to the year 1600, inclusive, for places on land or in its vicinity, for which the year of observation is known or for which it is possible to assign an approximate date. As the fact of the secular change of the magnetic declination did not become known until the next century, it was not customary to affix a date to an observation.[e] The sign ± in the table means that the date is approximate. The values obtained with sea compasses require careful scrutiny, as these compasses were frequently shifted to allow for the supposed variation or "error" of the needle. Thus, Robert Norman, instru-

---

[a] Actual mean was 11° 18' or nearly 11¼°, the quantity given by Gilbert in the "De Magnete." Both Norman and Borough persistently give 11° 15'. Gellibrand later recalculated Borough's observations, making allowance for atmospheric refraction, and deduced a mean value of 11° 16'. (See "Walker's Terrestrial and Cosmical Magnetism," 1866.)

[b] The table of values was obtained by Stevin from the cartographer P. Plancius, who is said to have entered them on a globe or a chart completed in 1592. Hence they refer to dates prior to 1592.

[c] From $\chi\alpha\lambda\upsilon\psi$ (genitive, $\chi\alpha\lambda\upsilon\beta os$), steel, and $\kappa\lambda\iota\nu\varepsilon\iota\nu$, to decline, hence, declination of the magnet.

[d] The term "variation" may have been derived from Guillen, who termed his instrument for determining it "brújula de variación." (See p. 26.)

[e] "And although this variation of the needle be found in Trauell to be divers and changeable, yet at any land or fixed place assigned it remaineth always one, still permanent and abyding." R. Norman, "The Newe Attractive," 1581.

ment maker, in 1581, says: "Of the common Sayling Compasses, I find heere (in Europa) five sundry sortes or sets"—according to the amount of correction allowed for by different makers. Thus, "by the Isle of Saint Michaell in the Acorres," he found "that the North poynt of the common compass, showeth the Pole very neere in that *Meridian*, but the bare Needle sheweth about 4 Degrees 50 Minutes to the Eastwards of the Pole."

It was not until the close of the sixteenth century that the "variation from the true north" came to be generally accepted as an actual fact of nature and not one to be accredited to the imperfection of the construction of the compass.

TABLE I.—*Earliest values of the magnetic declination up to 1600 for places on land or in its vicinity.*[a]

No.	Date	Place	Country	Latitude	Longitude	Magnetic Declination	Authority or observer
				° ′	° ′	° ′	
1	1436 (prior)	Rome	Italy	41 54 N	12 27 E	[b]5 E	L. A. Bauer from Compass Charts
2	1510±	do.	do.	41 54 N	12 27 E	6 E	Georg Hartmann
3	1518±	Bay of Guinea	Africa	....	....	(11½ E)	Piero di Giovanni d'Antonio di Dino
4	1520±	Vienna	Austria	48 15 N	16 21 E	4 E	Johann Georg Tannstetter (Rheticus)
5	1523 (?)	Landshut (?)	Germany	....	....	9 E	Petrus Apianus
6	1532±	Ingolstadt	do	....	....	10 30 E	Do.
7	1534	Dieppe	France	49 56 N	1 05 E	10 E	François or Crignon
8	1537	Florence	Italy	....	....	9 E	Mauro (Sphera volgare novamente tradotta. Venetia, 1537, 4°, fol. 53ª)
9	1538	Nuremburg	Germany	....	....	10 15 E	Georg Hartmann
10	1538, April	Lisbon	Portugal	38 42 N	9 08 W	7 30 E	João de Castro
11	1538, Aug. 10	Mozambique	Africa	15 02 S	40 46 E	6 45 E	Do.
12	1539±	Dantzig	Germany	....	....	13 E	Georg Joachim Rheticus
13	1541	Paris	France	48 52 N	2 20 E	7 E	Hieronymus Bellarmatus
14	1544±	Nuremburg	Germany	....	....	10 E	Georg Hartmann
15	1546±	Island Walcheren	Holland	....	....	9 E	Gerhard Mercator
16	1550	Paris	France	48 52 N	2 20 E	8 E	Orontius Finæus
17	1556, July 17	Petchora R. (mouth)	Russia	69 10 N	55 00 E	3 30 W	Stephen Borough
18	1556, July 27	Nova Zembla (S. coast)	do.	70 42 N	57 30 E	7 30 W	Do.
19	1556, Aug. 6	Vaigatch I. (coast)	do.	70 25 N	59 00 E	8 00 W	Do.
20	1557	Kholmogery	do.	64 25 N	41 50 E	5 10 E	Do.
21	1557, June 2	Dogsnose (2 miles on shore to northward)	do.	65 47 N	40 00 E	4 00 E	Do.
22	1557, June 16	Kola Peninsula	do.	66 59 N	39 30 E	3 30 E	Do.
23	1569	Böckstein	Austria	47 05 N	13 07 E	15 00 E	[Doppler's collection]
24	1575±5	St. Michael Island	Azores	37 00 N	25 00 W	4 50 E	R. Norman
25	1576, June	Gravesend	England	51 23 N	0 20 E	11 30 E	Frobisher
26	1576, June	Fair Island (SW. of)	Scotland	59 20 N	2 10 W	11 09 E	Do.

[a] Compiled from the following sources: No. 1 derived from Bianco's compass charts (see p. 21); 2-15, inclusive, from Hellmann and Wagner's collections (Journal "Terrestrial Magnetism," Vol. IV, p. 80, and Vol. VIII, p. 196); No. 24 from Norman's "The Newe Attractive," (see citation, p. 43; the date was approximately assigned). Nos. 25 and 33, from W. Borough's "Variation of the Compass," 1581. (Norman in his book also states that he found at London 11° 15′ by his own observation. Doubtless Borough and Norman made the London observation together.) The rest of the observations except No. 28 (see footnote e) are taken from Hansteen's "Magnetismus der Erde," and principally from van Bemmelen's valuable collections, "Abweichung der Magnetnadel," Batavia, 1899.

[b] It is a curious coincidence that this value agrees precisely with the one (5° E.) which had been for so long erroneously ascribed to Peregrinus, as having been observed by him in 1269. See p. 20.

TABLE I.—*Earliest values of the magnetic declination up to 1600 for places on land or in its vicinity*—Continued.

No.	Date.	Place.	Country.	Latitude.	Longitude.	Magnetic Declination.	Authority or observer.
				° ′	° ′	° ′	
27	1579	Bermejo Port	South America	50 25 S	75 00 E	0 00	P. Sarmiento de Gamboa
28	1579 (?)	Cape Mendocino (near)	California	39 00 N	124 00 W	a 9 00 E	Sir Francis Drake
29	1580, Apr. 17	Astrakhan	Russia	46 21 N	48 02 E	13 40 W	Chr. Bordugh
30	1580, June 11-16	Bildih	do.	40 25 N	49 30 E	10 40 W	Do.
31	1580, Oct. 4	Derbent	do.	42 05 N	48 15 E	11 00 W	Do.
32	1580, Oct. 16	London	England	51 31 N	0 08 E	11 15 E	W. Boroughs and R. Norman
33	1580	Paris	France	48 52 N	2 20 E	11 30 E	Severtius
34	1581 (before)	Vaigatch Island	Russia	70 ± N	58 ± E	7 00 W	W. Boroughs
35	1587, Apr.	Maipo	South America	34 00 S	71 39 W	2 30 W	Cavendish
36	1587, May 25	Puna	do.	2 45 S	80 00 W	2 00 E	Do.
37	1587, Aug.	Mauranilla	Mexico	18 15 N	104 00 W	2 00 E	Do.
38	1587	Cape Corientes	do.	20 45 N	106 00 W	2 00 E	Do.
39	1587	Cape San Lucas (near)	do.	22 55 N	111 56 W	3 00 E	Do.
40	1587, June 30	Greenland, E. coast		72 10 N	56 00 W	28 00 W	Davis
41	1587, July 23	Cumberland Bay, NW. end		67 00 N	67 30 W	30 00 W	Do.
42	1589, Aug. 14	Santa Cruz (Flores)	Azores	39 50 N	30 40 W	b 4 00 W	Edward Wright
43	1589, Sept. 13	Fayal, in the town	do.	38 50 N	27 40 W	b 1 30 E	Do.
44	1589, Sept. 22	do.	do.	38 50 N	27 40 W	b 4 40 E	Do.
45	1589, Sept. 23	do.	do.	38 50 N	27 40 W	b 3 50 E	Do.
46	1589, Nov. 12	NE. of Cape Finisterre	Spain.	44 25 N	10 00 W	c 7 00 E	Do.
47	1594	Off Cape St. Vincent	do.	37 05 N	9 10 W	5 15 E	Robert Dudley
48	1595, Jan.	Off Cape Barbas	Africa	22 00 N	17 00 W	3 00 E	Do.
49	1595, Jan.	Off Cape Roxo	Porto Rico	17 54 N	67 05 W	3 00 W	Do.
50	1595, Aug. 4	Bay Aguada de Sambras (Mossel Bay?)	Africa	34 10 S	22 00 E	0 00	Corn. Houtman
51	1595, Sept. 3	Off Cape San Roman	Madagascar	25 30 S	46 50 E	13 00 W	Do.
52	1596, June 22	Entrance Sunda Sts.		6 00 S	104 20 E	4 00 W	Do.
53	1596, June 9	Bear Island (Cherry)		74 10 N	16 00 E	13 00 E	Willem Barentsz
54	1596, June 23	Hinlopen Strait	Spitzbergen	79 40 N	17 00 E	16 00 W	Do.
55	1596, July 21	Nova Zembla, Cross I.	Russia	76 45 N	59 00 E	26 00 W	Do.
56	1596, July 31	do.	do.	76 45 N	59 00 E	17 00 W	Do.
57	1596	Nova Zembla, Langeneus	do.	73 40 N	53 30 E	25 00 W	Do.
58	1596	Vaigatch Island	do.	69 10 N	61 10 E	24 30 W	Do.
59	1596	Williams Island	do.	75 50 N	58 30 E	33 00 W	Do.
60	1596	Yshoek	do.	76 55 N	67 30 E	27 00 W	Do.
61	1596	Nova Zembla	do.	76 07 N	68 34 E	26 00 W	Do.
62	1596-99	Graz	Austria	47 07 N	15 25 E	6 00 W	J. Kepler
63	1597, Feb.	Bali Strait, east end of	Java	8 30 S	114 50 E	d 2 00 W	Corn. Houtman
64	1597, Apr. 24	Africa (SE. coast)		32 30 S	28 50 E	5 00 W	Do.
65	1597, May 4	Off Cape of Good Hope		34 50 S	18 20 E	0 30 E	Do.
66	1597, Aug. 11	Off Egmont	Holland, coast	52 30 N	4 20 E	15 00 E	Do.
67	1598, June 28	Off Martin Vaz I.		20 38 S	31 13 W	11 10 E	Van Neck
68	1598, Sept. 28	Off Mauritius Island		20 27 S	67 30 E	22 15 W	Do.
69	1598, Dec. 31	Off Bantam	Java	6 00 S	106 10 E	5 10 W	Do.

a This value is given on a map by R. Dudley in the "Arcano del Mare," and preserved by Petrus Koerius, dated 1646, showing the coast of New Albion, discovered by Sir F. Drake in 1579. Narrative and critical history of America, Justin Winsor, vol. 2, Boston and New York, 1886.

b These observations, according to Hansteen, were made by Wright with W. Boroughs' compass described in B.'s book.

c This value is given by Hansteen in one place as 7° 40′, in another as 7° 04′; Van Bemmelen apparently rounds off the value to 7°.

d Not quite 2°.

TABLE I.—*Earliest values of the magnetic declination up to 1600 for places on land or in its vicinity*—Concluded.

No.	Date.	Place.	Country.	Latitude.	Longitude.	Magnetic Declination.	Authority or observer.
				° '	° '	° '	
70	1599, Feb. 9	Off Arosbaya	Madura Island	7 00 S	112 50 E	2 30 W	Van Neck.
71	1599, Apr. 3	Amboina, west end		3 26 S	128 30 E	3 10 E	Do.
72	1599, Apr. 19	Off Ternate and Tidore		1 02 N	127· 20 E	3 10 E	Do.
73	1600, May 7	Off St. Helena Island		15 55 S	5 43 W	7 38 E	Do.
74	1600, May 22	In bay, I. Ste. Marie	Madagascar	15 40 S	47 30 E	16 30 W	Wilkens
75	1600, July 13	Off Maldive Islands	Indian Ocean	2 00 N	73 00 E	15 00 W	Do.
76	1600	Between Buru and Amboina	Dutch E. I.	3 45 S	127 30 E	3 00 E	Do.
77	1600, Sept.	Off Bantam	Java	6 00 S	106 10 E	5 00 W	Do.
78	1600	Constantinople	Turkey	41 01 N	28 50 E	0 00	
79	1600 (before)	Antwerp	Belgium	51 13 N	4 24 E	9 00 E	
80	1600	Konigsberg	Prussia	54 42 N	20 26 E	0 00	
81	1600 (before)	Plymouth	England	50 26 N	4 19 W	13 24 E	
82	1600, Sept. 26	Cape San Sebastian	Madagascar	12 42 S	47 40 E	16 00 W	J. Laukester

Glancing over these values, it will be seen that in the sixteenth century the needle pointed east of north over the greater part of Europe, whereas now it as persistently points west, except in the eastern part. *Cf.* the charts of lines of equal magnetic declination for 1500 (Fig. 4) and 1600 (Fig. 17).

DISCOVERY OF THE MAGNETIC INCLINATION.

The year 1581 is memorable as having produced the first two works treating distinctively of the earth's magnetism. The first, that of Robert Norman, entitled "The Newe Attractive,"[a] heralded to the world an entirely new fact about the magnetic needle—"a newe discovered secret and subtill propertie concernyng the Declinyng of the Needle, touched therewith under the plaine of the Horizon." *This discovery of the dip of the needle below the horizon was made in 1576 by Norman, a practical seaman, or "hydrographer," as he styles himself, and an instrument maker. Thus the second element of the earth's magnetism came to light and gave another incentive for magnetic measurements.* In Chapter III of his quaint and exceedingly rare book he relates "by what meanes the rare and strange declining of the Needle, from the plaine of the horizon was first found."

"Hauing made many and diuers compaffes, and ufing alwaies to finifh and end them before I touched the needle, I found continually, that after I had touched the yrons with the Stone, that prefently the north point thereof would bend or *Decline* downwards under the Horizon in fome quantitie: infomuch that to the Flie of the compaffe, which before was made equall, I was ftill conftrained to put fome fmall peece of waxe in the South part thereof, to counterpoife this *Declining*, and to make it equall againe.

"Which effect having many times paffed my hands without any great regard thereunto, as ignorant of any fuch propertie in the Stone, and not before hauing heard nor

---
[a] Principal parts reproduced in facsimile in Hellmann's reprints, "Rara Magnetica," Berlin, 1898.

read of any fuch matter: It chaunced at length that there came to my hands an Inftrument to bee made, with a Needle of fixe inches long, which needle after I had pollifhed, cut off at Juft length, and made it to ftand levell upon the pinne, fo that nothing refted but onely the touching of it with the stone: when I had touched the fame, prefently the north part thereof *Declined* downe in fuch fort, that beeing conftrayned to cut away fome of that part, to make it equall againe, in the end I cut it too fhort, and fo fpoyled the needle wherein I had taken fo much paynes.

"Hereby beeing ftroken in fome choller, I applyed my felf to feeke further into this effect, and making certayne learned and expert men (my friends) acquainted in this matter, they advifed me to frame fome Inftrument, to make fome exact tryal, how much the needle touched with the Stone would *Decline*, or what greateft Angle it would make with thee plaine of the Horizon. Whereupon I made diligent proofes: the manner whereof is fhewed in the Chapter following."

Chapter IV next tells "how to finde the greatest Declining of the Needle, under the Horizon":

"Take a fmall Needle of Steele wier, of five or fixe inches long, the fmaller and the finer mettall the better, and in the middle thereof (croffe the fame) by the beft means you can, fixe as it were a fmall Axeltree of yron or braffe, of an inch long, or thereabout, and make the ends thereof very fharpe, whereupon the Needle may hang levell, and play at his pleafure.

"Then provide a round plaine Inftrument like an Aftrolobe, to be divided exactly into 360 partes, whofe diameter muft be the length of the Needle, or thereabout, and the fame inftrument to bee placed uppon a foot of convenient height, with a plumme line to fette it perpendicular.

"Then in the Center of the fame Instrument place a peece of Glaffe hollowed, and againft the fame Center uppon fome place of Braffe that may be fixed upon the foot of the Inftrument, fit another peece of Glaffe, in fuch forte that the fharpe endes of the Axeltree beeing borne in thefe two Glaffes, the Needle may play freely at his pleafure, according to the ftanding of the Inftrument.

"And the Needle muft be fo perfected, that it may hang upon his Axeltree both ends levell with the Horizon, or being turned, may ftand and remaine at any place that it fhall be fette: which being done, touch the faide Needle with the *Magnes* ftone, and fet the Inftrument perpendicular by the plumme line, and turne the edge of the Inftrument South and North, fo as the Needle may ftand duley according to the *Variation* of the place: which *Variation* the Needle of his owne propertie would fhew, were it not that he is conftrained to the contrarie by the Axeltree.

"Then fhall you fee the *Declination* of the North point of the touched Needle, which for this Citie of London, I finde by exact obfervation to be about 71 degrees 50 minutes. This forme of the inftrument heere defcribed with the manner of the declination, I have heere placed that it may be the eafier conceived."

He next proves by experiment and weighings that it is not want of balance of needle nor the rubbing of it with the loadstone that makes this "declining of the needle."

One can not but admire the painstaking and conscientious labors of Norman and the precision with which he set out to determine the amount of "declining." It will be noted that he explicitly states that the angle must be determined with the instrument standing "duley according to the Variation of the place"—that is, in the magnetic meridian. It is curious, however, that he should call this the "greatest declining," whereas in the plane of the magnetic meridian the declining is really *the least*, the angle increasing as the instrument is turned away from the magnetic meridian and reaching its maximum amount of 90° in a magnetic east and west plane. How exact his observation of 71° 50′ is can not be judged in the absence of further details.

From the letter, cited on page 26, which the famous vicar of Nuremburg, Georg Hartmann, wrote March 4, 1544, to Count Albert of Prussia, it is apparent that he had already become aware of the dipping of the north end of the needle. He says: "Besides, I find this also in the magnet, that it not only turns from the north and deflects to the east about 9° more or less, as I have reported, but it points downward. This may be proved as follows: I make a needle a finger long, which stands horizontally on a pointed pivot, so that it nowhere inclines toward the earth but stands horizontal on both sides. But as soon as I stroke one of the ends (with the lodestone), it matters not which end it be, then the needle no longer stands horizontal, but points downward some 9° more or less. The reason why this happens I was not able to indicate to His Royal Majesty."

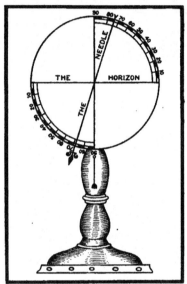

FIG. 6.—First Dip Circle (Norman's 1576).

Hartmann's letter was not published until it was rescued from oblivion in the third decade of the nineteenth century, and its contents do not appear to have been known to Norman or to any of the writers of that period. It was recently republished in facsimile by Hellmann in his "Rara Magnetica." Hartmann did not mount his needle in such a manner as to show the precise amount of dip, as did Norman, but simply observed the dip of the north end of a *compass* needle, mounted as ordinarily, on a pivot, so that instead of getting about 65°, as he ought to have done, he only found 9°. As is well known the dip of the north end of a compass is nowadays usually overcome in the northern magnetic hemisphere by a sliding brass weight or ring on the south end. *Accordingly, the principal credit for the discovery of the magnetic dip must undoubtedly be assigned to Norman.*[a]

It is a keen pleasure to peruse Norman's book, which was so popular that it was

---

[a] It has also been claimed that reference to the dip of the needle is made in Fortunius Affaytatus's book, "Physicae et astronomicae," published in 1549, but this does not appear to be the case.

republished four times (1585, 1596, 1614, and again in 1720, bound with Whiston's treatise), and note the admirable and modest manner in which he relates his experiments and discoveries, differing greatly in this respect from Gilbert, who, in his great work (1600), vehemently abuses almost every writer on magnetism and rarely credits anyone with the facts previously discovered.

Norman must clearly be given credit for being the first to divine that the point or source of power which the needle respects is in the earth and not in the heavens, as had been generally supposed before his time. He says:

"And by the *Declining* of the Needle, is alfo proved, that the point *Refpective*, is rather in the earth than in the Heavens, as fome have imagined; and the greateft reafon why they fo thought (as I judge) was becaufe they never were acquaynted with this *Declining* in the Needle, which doubtleffe if *Martin Curtes* had known, he would not have judged the *Attractive* point to have been in the Heavens, or without them, but rather in the earth."

Note also this remarkable sentence: "And surely I am of opinion, that if this Vertue could by any means be made vifible to the Eye of man, it would be found in in a sphericall forme, extending rounde about the Stone in great Compaffe, and the dead bodie of the Stone in the middle thereof. Whose center is the center of his aforefaid Vertue. And this I have partly prooved and made Vifible to be seene in some manner, and God sparing mee life, I will herein make further Experience and that not curioufly, but in the Feare of God, as neare as he shall give mee grace, and meane to annex the same unto a Booke of Navigation, which I have had long in hand."

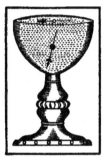

FIG. 7.

This is undoubtedly the source from which Gilbert got his idea of the "orbs virtutis"—the circular orb of virtue surrounding the globular lodestone. In fact, Gilbert in no way improves on Norman's idea but adopts it bodily. Some writers have extravagantly asserted that Gilbert anticipated Faraday's conception of the field of force surrounding a magnet.

Norman also proves experimentally that the attraction exerted on the magnet does not produce motion of translation but simply that of rotation (of the compass needle and of the dip needle).[a] His figure illustrating the experiment is herewith (Fig. 7) reproduced (half size).

[a] In experiments with the terrella the needle is attracted obliquely or directly toward the globe with a very perceptible force. This is because the length of the needle is so considerable in proportion to the diameter of the globe that the magnetic forces on the two ends are not equal and parallel. But the length of the longest of mariner's compass needles is not more than about $\frac{1}{10,000,000}$ and the length of the largest bar magnet that has ever been suspended so as to show by its movements any motive tendency it may experience from the force of terrestrial magnetism is not more than $\frac{1}{10,000,000}$ of the Earth's diameter, and therefore magnetic needles or bar magnets experimented on in any part of the world experience no sensible attraction toward or repulsion from the Earth and show only a directional tendency according to which a certain line of the magnet, called its magnetic axis, takes the direction of the curved lines of force. ("Terrestrial magnetism and the mariner's compass," by Sir W. Thomson (Lord Kelvin) in Popular Lectures and Addresses, Vol. III, Navigation, pp. 228-337).

THE EARTH, A GREAT MAGNET.

GILBERT'S "DE MAGNETE."

The year 1600 is generally regarded as a memorable one in the history of the sciences of magnetism and electricity, for in this year appeared Dr. William Gilbert's famous work "De Magnete," published at London, dedicated in his prefatory remarks to the "True philosophers, ingenuous minds who not only in books but in things themselves look for knowledge," and treating in five books or sections of the properties of magnetic bodies and of the "great magnet, the Earth." It was republished in Latin at Stettin (Sedini) in 1628 and 1633 by Wolfgang Lochmann, reprinted in 1892 in facsimile (photozincograph reproduction of 1600 edition) by Mayer and Müller, of Berlin, and translated into English for the first time by P. Fleury Mottelay,[a] and more recently under the auspices of the Gilbert Club.[b]

William Gilberd, or as more usually written Gilbert, was born in the year 1540 in Holy Trinity Parish at Colchester, England, being the eldest of five sons of Jerome Gilbert, at one time recorder. Matriculating at the age of 18 at St. John's College, Cambridge, he in due course took the degree of B. A.; he also became a Symson Fellow in 1561, an M. A. in 1564, and during the two years following was mathematical examiner of his college. He next studied medicine, reaching his doctorate and a senior fellowship in 1569, when he terminated his eleven years' connection with the university, after which he spent four years on the Continent.

Upon his return to London he practiced as a physician for thirty years with "great success and renown," and was made a Fellow of the Royal College of Physicians, later censor, then treasurer, next consilarius, and finally, in 1600, president of the college. In the same year Queen Elizabeth appointed him one of her body physicians and settled upon him a pension to enable him to prosecute his scientific researches. After her death Gilbert was continued in his office by James I. He died in November, 1603, and was buried in Trinity Church, Colchester. His books, papers, and collections, bequeathed to the Royal College of Physicians, were unfortunately destroyed in the "great fire."

It is not known how Gilbert, a successful physician, was led to devote himself so zealously and so unremittingly to the study of magnetism. He says "There is naught in these books (De Magnete) that has not been investigated and again and again done and repeated under our eyes." Herein consists the chief value of the work—that nearly every conclusion drawn rests on experiment made over and over again under slightly varying conditions, for, as he says, "stronger reasons are obtained from sure experiments and demonstrated arguments than probable conjectures, and the opinions

---

[a] Published in 1893 by Quaritch, of London, and Wiley & Sons, of New York.

[b] President of the Club, Lord Kelvin. The translation was prepared from the original edition of 1600 by a Committee of the Club formed for this purpose in 1889, which finished its labors in 1900. The printing was undertaken in 1901 at the Chiswick Press by Messrs. C. Whittingham & Co., the edition being unfortunately limited to 250 copies. Prof. Sylvanus P. Thompson, one of the secretaries of the Club who took a most active part in the translation, has issued at his own expense his most valuable and useful commentaries, entitled: "Notes on the De Magnete of Dr. William Gilbert," privately printed, London, 1901. As the Gilbert Club's translation is not yet at hand, the quotations given above are according to Mottelay.

of philosophical speculators of the common sort." It is said that he spent £5,000 on his experiments, "examining very many matters taken out of the lofty mountains, of the depths of the seas, or deepest caverns, or hidden mines," in order to discover the true substance of the Earth and of magnetic forces.

The De Magnete was the most complete summary of the properties of magnetic bodies up to 1600. One reading this work alone, however, must by no means infer that all the properties and laws set forth were discovered by Gilbert, for he very rarely gives credit to any previous discoverer. He frequently exhibits intolerance and lack of appreciation of the work of his predecessors, and like his experiments, repeats his vituperations and assertions over and over again, so that one is unconsciously led to believe that all previous work had resulted in very little of real value.

Doubtless the fact that he thoroughly tested anew everything he had heard regarding magnetic substances, and accepted nothing on faith led him to regard all as his own and thus prevented him from giving credit where credit was rightfully due. The weak points of others, however, he never fails to expose and ridicule.

Gilbert terms the end of the lodestone or needle which points to the north, the south pole, and the opposite end, the north pole, for similar reasons to those already set forth. And by reiterating over and over Gilbert would apparently desire to convey the impression that he was the first to recognize the fact that the magnetism residing in the north-pointing end of a magnetic needle is of an opposite kind to that at the Earth's north magnetic pole, although this fact was clearly recognized by many writers previously, beginning with Peregrinus in 1269. Gilbert must be simply credited with proposing to designate, because of the fact stated, the north-seeking end of the needle, the south pole—a proposal which, by the way, has not been accepted by modern writers.

One can not fail, however, to recognize that Gilbert did a most useful piece of work in so carefully scrutinizing, weighing, and summarizing in suggestive and descriptive language all knowledge of magnetic properties. *As a work on magnetism and electricity, Gilbert's De Magnete is still a standard one; as a work on terrestrial magnetism, however, it was weak even for its time, its conclusions and deductions having all been discredited with the exception of one, the truth of which he got right more by chance than by philosophical reasoning, viz, that the "Earth itself is a great magnet."*

As said, Gilbert's work as a treatise on terrestrial magnetism was by no means equal to his work on the general properties of magnetic bodies. When he came to theorize on the "Earth as a magnet" he forgot his own injunction to philosophers who but dream and speculate from books, saying that they "must be aroused and taught the uses of things, the dealing with things; they must be made to quit the sort of learning that comes only from books,[a] and that rests only on vain arguments from probability and upon conjectures."

Although he is credited as having determined a dip of 72° at London, and by Kircher as having found the declination to be 6°[b] at London, his work contains nothing to lead one to suppose that he obtained the declination and dip himself. He repeatedly points out the errors of observations by others, but makes no attempt whatsoever to

---
[a] Gilbert might have added: "and mere laboratory experiments."
[b] In 1580 the declination at Limehouse, London, was 11¼° E., and in 1600 about 10° E.

test by actual observation at various places the deductions drawn from his terrella, or spherical lodestone, and directly applied to the Earth. It is claimed that the chapter on methods for finding the "variation of the compass" was written by his friend Edward Wright, a practical navigator. His book does not even contain a systematic collection of all observations up to his time, such as that made, for example, by Plancius and published in Stevin's work the year before. Had Gilbert been equally as zealous in observing the terrestrial magnetic elements as he was in his laboratory experiments, he might have stumbled on a fact—the secular change of the magnetic declination—which would doubtless have shaken him, to some extent at least, in his belief that the "Earth was a great lodestone;" for one of the fixed and necessary postulates of his theory was the constancy of the magnetic declination at any place.

Gilbert reached his conclusion that the "Earth is a great magnet," *i. e.*, that its "magnetic virtue" comes from within the Earth and not from the heavens above, solely by analogy between the Earth and a globular lodestone which he termed a "terrella," and which he had had expressly made for his experiments to represent the Earth on a miniature scale. The reasoning whereby he was led to the conclusion (Book I, Chapter XVII) that the "terrestrial globe is magnetic and is a lodestone," upon which his fame largely rests, would not be accepted to-day, and, in fact, was not accepted by writers after the discovery of electro-magnetism. The problem was not definitely settled until Gauss, in 1838, attacked it analytically, with the aid of the observations accumulated up to his time, and showed that the Earth derives its permanent magnetism almost entirely from sources residing within its own crust, and not, for example, from any system of electric currents circulating around the Earth in the upper regions.

The recent researches of Dr. Schmidt, of Gotha, have confirmed Gauss's conclusion. He finds that about 95 per cent of the Earth's magnetic force is to be referred to causes within its crust and the remainder to electric currents either circulating around the Earth in the upper regions or passing from the air into the earth, and *vice versa*. Some of the periodic and spasmodic variations of the Earth's magnetism, such as the diurnal variation, annual variation (not secular change) and magnetic perturbations, according to recent researches by Schuster. von Bezold, Schmidt, Schwalbe, and others, would apparently have to be ascribed to electric currents in the upper regions.

If the way the compass points at various places on the Earth constituted the entire knowledge on the subject, it would be impossible to say whether the compass approximately points northward because of magnetism (or electric currents) within the Earth or external to it. There are, undoubtedly, in the Earth's crust large masses of magnetized and magnetizable substances, as Gilbert inferred from the specimens collected from many parts of the Earth, but modern researches would indicate that the chief source of the Earth's magnetism is not to be referred to permanently magnetized substances, but doubtless to a system of electric currents embedded deep within the interior of the Earth and connected in some manner with the Earth's rotation. In order to make the compass point northward, the electric currents would have to circulate in the interior from east to west, in accordance to the well-known rule of Ampère governing the deflection of a magnetic needle by an electric current. The compass can be made to point north equally as well, however, by electric currents circulating around the Earth in the upper regions in the contrary direction, viz, from west to east. Therefore with the aid of the compass needle alone it could not be determined whether the currents are inside or outside the Earth.

The dip needle will determine this. The fact that the same end of the compass which points north likewise dips downward in the northern magnetic hemisphere requires, as can be easily shown by applying Ampère's rule, that the electric currents circulate from east to west, and hence, in accordance with the evidence furnished by the compass and the dip needle, the currents must be in the interior of the Earth.

Now, while Gilbert had at his command a general knowledge of the pointing of the compass needle over the regions then traversed, he only had one dip observation to work with—that made by Norman at London in 1576, and doubtless verified by himself. He does not appear to appreciate that it is the salient feature of the dip needle which reveals the fact that the "Earth itself is a great magnet." The citation from Norman's book, page 33, shows that by the discovery of the dip Norman had already inferred that the "point respective" which the needle heeds "is rather in the Earth than in the Heavens," and Gilbert in no wise improves upon or adds anything to Norman's reasoning.

To Gilbert the Earth was but a great round lodestone. It had poles and an equator, just as the terrella had its magnetic poles and a natural line or magnetic equator halfway between; it took a definite position in space, just as the terrella did with reference to the Earth; it had its diurnal motion[a] and revolution, just as the terrella had when floated in a bowl of water and brought under the action of the Earth's force; it contained in abundance the very lodestone substance which possessed this remarkable "magnetical virtue;" it magnetized substances just as did the lodestone; it, like the lodestone, attracted bodies to itself (Gilbert regarded gravity and magnetism as identical); therefore, like the lodestone, it was a magnet. All of this reasoning would equally apply for the magnetic effects due to an outside electric field, but in Gilbert's time, though he could distinguish between them, the mutual relationship between electric and magnetic phenomena had not been discovered. He only knew of permanent magnets such as are exhibited in lodestones and artificially made magnets.

According to Gilbert's theory, the Earth's magnetic poles were coincident with the rotation poles; in fact, he regarded the Earth's rotation as due to magnetic action. The compass, therefore, if it had not been "perverted" in its direction by the attracting influence of the continents, as he thought, would accordingly point true north and south. He persistently regarded the magnetic declination, or, as he termed it, the "variation," as a "sort of perturbation and depravation of the true direction." The Germans, in their term of "*missweisung*," *misdirection*, convey a similar idea. It never entered Gilbert's mind to consider the "variation" as due, in whole or in part, to noncoincidence of magnetic poles and rotation poles, for, were that true, his theory of the Earth as a great lodestone would have fallen to the ground.

He accordingly seeks another explanation, viz, that the "variation" is due to the fact that the elevated and massive parts of the Earth (continents) are more strongly magnetic, and the waters of the globe less so; hence the needle is drawn toward

---

[a] Gilbert has the credit of being one of the earliest and most ardent advocates in England of Copernicus's theory of the diurnal rotation of the Earth. His magnetic theory of the Earth was in fact largely, if not entirely, advanced in order to furnish a cause for this diurnal rotation.

the continents. He ignominiously fails, however, in this explanation, and apparently ignores facts, undoubtedly known to him, which would have contradicted his theory. He lays himself open here to the same kind of criticism which he so unsparingly heaped upon others.

Apparently aware of the fact that the dip of the needle at London did not correspond to what it ought to have been on the theory that the magnetic poles are at the geographical poles, he speaks of a "variation of the dip," and ascribes this to the same cause as the "variation of the compass." Aware that in the dip the same kind of variations, though not of the same degree as in the magnetic declination, might be expected, he nevertheless proposes a method for determining latitude by means of the dip needle. And yet he ridicules those who had proposed to determine the longitude by means of the magnetic declination.[a]

To conclude, while it must be conceded that Gilbert made the first serious attempt to correlate the magnetic phenomena of the Earth and to construct a theory, his actual and real contributions to the subject of the Earth's magnetism are by no means of that brilliancy and luster which is generally supposed, and which mark his other works, his failures being due in a large degree to his not following his own advice to philosophers, "to leave their books and go out and deal with things." In the writer's estimation, Norman's little work should be given a higher rank as a real and valuable contribution to our advancement of the knowledge of the Earth's magnetism than that part of Gilbert's book dealing with terrestrial magnetism.

THE VARIATIONS OF THE EARTH'S MAGNETISM.

DISCOVERY OF THE SECULAR CHANGE OF THE MAGNETIC DECLINATION.

The only contribution of great value to the science of terrestrial magnetism in the seventeenth century was the discovery of the secular change of the magnetic declination by Gellibrand in 1634.[b] Hitherto it had been supposed that the magnetic declination, though varying from place to place, was fixed and invariable at any one place, except that "by the break up of a continent," as Gilbert put it, it might suffer a change. But now an entirely new and most important fact came to light, showing indisputably that

---

[a] The suggestion of determining the longitude at sea by means of the magnetic declination started with Columbus and served to stimulate the making of magnetic observations until the close of the eighteenth century. In 1720 William Whiston, the translator of Josephus, revived Gilbert's idea of using the dip and accordingly supplied certain mariners with dip circles. Thus some notable contributions to terrestrial magnetism were obtained. The earliest dip observation in the United States is that made at Boston in 1722 with a dip circle supplied to Capt. Othniel Beal by Whiston.

[b] Some of the principal writers on magnetism and terrestrial magnetism of the seventeenth century besides Gellibrand were: Barlowe, in whose book, Magnetical Advertisements, 1616, the word "magnetism" as a noun, according to Prof. Silvanus P. Thompson, appears for the first time; Mark Ridley, Bacon, Galileo, Nicolaus Cabaeus, whose Philosophia Magnetica, Ferrara, 1629, the first Italian treatise on the magnet, contains an improvement of Gilbert's picture of the lines of force around a magnet; Kepler, Athanasius Kircher (Jesuit and an opponent of the Copernican theory), who in his works collected all values of the magnetic declination known to him; Descartes, Porta, von Guericke, Hooke, and Bond, who made a special study of the subject of the secular change in the dip, using the word "inclination" to denote the dip in place of the word "declination," which, as will be recalled, Norman had employed.

the earth's magnetism suffers mighty changes in the course of time. Hence it now became necessary to note not only the place but also the time when an observation of the magnetic declination was made. The compass had by this time come into general use, not only as an instrument, as Gilbert said, "beneficial, salutary, and fortunate for seamen, showing the way to safety and to port," but also for the purpose of running out lines on the earth's surface (land surveys) and in mines, and for the orientation of buildings. To retrace these lines anew at some subsequent period required a consideration of the newly discovered fact. No wonder this truth was fought, disputed, and doubted for some time.

Henry Gellibrand was a professor of mathematics at Gresham College. He made a careful determination of the pointing of the compass on June 12, 1634, at Diepford, or Deptford, about 3 miles southeast of London Bridge, and found 4° 6′ east. Now, Edmund Gunter, another mathematician of Gresham College, had found on June 13, 1622, 5° 56½′ east, and, as will be recalled, Borough and Norman had found in 1580, 11° 15′ east. Clearly, therefore, the magnetic declination had suffered considerable change since 1580. Gellibrand repeated his observations, next examined carefully the observations which Borough had published, and although he found that Borough had neglected to take into account atmospheric refraction in his calculations, nevertheless he got practically the same amount as Borough had given.

He announced his discovery in a book,[a] now exceedingly scarce, entitled "A Discourse Mathematical on the Variation of the Magneticall Needle, together with its admirable Diminution lately discovered." London, 1635. He says: "Thus (hitherto according to the Tenents of all our Magneticall Philosophers) we have supposed the variation of all particular places to continue one and the same; so that when a seaman shall happily return to a place where formerly he found the same variation, he may hence conclude 'he is in the same former *longitude*.' For it is the Assertion of *Mr. Dr. Gilbert: Variatio uniuscujusq; Loci constans est*, that is to say the same place doth alwayes retaine the same variation. Neither hath this Assertion (for aught I ever heard) been questioned by any man. But most diligent magneticall observations have plainely offered violence to the same, and proved the contrary, namely that the variation is accompanied with a variation."

He republishes the observations of 1580 and 1622, along with his own, in order to furnish all necessary evidence, and says:

"If any affected with magneticall Philosophy shall yet desire to see an experiment made for their owne particular satisfaction, where I may prevaile, I would advise them to pitch a faire stone parallel to the Horizon there to rest immoveably, and having a Needle of a convenient length strongly touch't by a vigorous Magnet to draw a Magneticall Meridian thereby, and yearly to examine by the application of the same (well preserved from the ayre and rust, its greatest enemies) whether time will produce the like alterations."

Most commendably and remarkably for his times, Gellibrand refrains from "entering into a dispute [speculation] concerning the source of this sensible diminution, whether it may be imputed to the magnet or the Earth, or both," but says it "must be all

---

[a] Reprinted in facsimile by Hellmann; Asher & Co., Berlin. Hellmann used a copy loaned him by the late Latimer Clark, whose exceedingly valuable library has come into the possession of the American Institute of Electrical Engineers, headquarters, New York.

left to future times to discover, this Invention being but newly presented to the world in its infancy."

The following sentence, taken from the article on the compass in such an authoritative work as the Encyclopædia Britannica, ninth edition, illustrates the great confusion caused by the misuse of the word "variation."

"The discovery of the variation of declination was made by Stephen Burrows when voyaging between the north cape of Finmark and Vaigatch (Vaygates), and was afterward determined by Gellibrand, professor of geometry at Gresham College."

In the first case the author means simply the change in the magnetic declination with geographical position, i. e., the geographical variation of the magnetic declination, whereas when referring to the discovery of Gellibrand, the slow variation taking place with the lapse of time, viz, the secular change, is meant. The author has thus used the word "variation" in the same sentence with two totally different meanings, preventing one thereby from getting a proper idea as to the precise facts involved. Besides, the geographical variation of the declination had been discovered in the century previous to that of Burrows's time, as already stated, by Columbus.

Nearly three centuries have passed since Gellibrand's discovery was made known, and although observations have been multiplied and some of the best minds have given their undivided attention to this most striking fact of the Earth's magnetism, the riddle is still unsolved. Innumerable theories have been advanced, the difficulty not being in finding a cause, but to tell which one among the many assignable ones is *the one*. While observations of declination for three centuries are at hand, those of dip are not so numerous and those of the intensity of the magnetic force are still more scarce, beginning only since the third decade of the last century. *Both the dip and intensity undergo secular change in the same manner as the declination.* The definite solution of this great and important problem of the Earth's physics requires a full and accurate knowledge of the changes in the three magnetic elements named. The prospects at present are fair that the secular change of the Earth's magnetism is to be referred, primarily, to the effect of secondary electric currents generated within the Earth by its rotation around an axis not coincident with its magnetic axis.

THE CHARACTERISTICS OF THE SECULAR CHANGE.

The secular change has received the closest attention in the United States, largely for practical reasons, as in all of the older States the original land surveys were referred to compass lines. The retracing of the "metes and bounds" at subsequent periods called for a knowledge of the amount of change in the compass bearing during the elapsed interval. To meet the demand for knowledge of this kind, C. A. Schott, who directed the magnetic work of the Coast and Geodetic Survey for nearly a half century, undertook a thorough and systematic collection of all known values of the magnetic declination in the United States and vicinity, resulting in a collection as yet unequaled in any other country.

It is a lamentable fact that such collections have not been undertaken for European countries, where in many instances the records go back to the sixteenth century. Knowledge of the manner and rate of progression of a particular phase of the secular change from place to place would be materially increased thereby.

The following table[a] exhibits how the declination has changed at various places:

TABLE II.—*Showing the secular change in the magnetic declination at various places.*

Year	Northern Hemisphere						Southern Hemisphere			
	London	Paris	Rome	Manila	San Francisco	Baltimore	Rio de Janeiro	Ascension Island	St. Helena Island	Cape Town
1540	7.2(?) E	8.2 E	10.47 E							
1560	9.6(?) E	9.3 E	11.61 E							
1580	10.93 E	9.6 E	11.41 E							
1600	10.13 E	8.8 E	9.88 E							
1620	7.26 E	6.9 E	7.29 E							
1640	3.27 E	4.42 E	3.86 E			5.3 W				
1660	0.59 W	0.86 E	0.01 W			6.0 W				
1680	3.89 W	3.47 W	4.01 W			6.1 W				
1700	7.08 W	7.99 W	7.77 W			5.5 W				
1720	10.97 W	12.27 W	11.02 W			4.5 W				
1740	15.30 W	15.83 W	13.63 W			3.2 W				
1760	19.57 W	18.76 W	15.51 W			1.95 W	8.6 E	8.4 W	11.70 W	20.5 W
1780	22.65 W	20.87 W	16.64 W		12.6 E	1.03 W	7.2 E	11.6 W	14.59 W	23.2 W
1800	24.07 W	22.12 W	17.06 W	0.08 E	13.6 E	0.66 W	5.5 E	14.0 W	17.51 W	25.4 W
1820	24.09 W	22.40 W	16.77 W	0.14 E	14.6 E	0.93 W	3.6 E	16.4 W	20.01 W	27.2 W
1840	23.22 W	21.38 W	15.84 W	0.27 E	15.43 E	1.77 W	1.2 E	18.8 W	22.00 W	28.8 W
1860	21.55 W	19.54 W	14.23 W	0.45 E	16.11 E	2.99 W	1.4 W	21.4 W	23.41 W	29.7 W
1880	18.73 W	16.76 W	11.77 W	0.69 E	16.57 E	4.30 W	4.3 W	22.9 W	24.11 W	29.6 W
1890	17.57 W	15.16 W	10.57 W	0.83 E	16.64 E	4.89 W	6.1 W	23.0 W	24.21 W	29.2 W
1900	16.5 W	14.6 W		0.97 E	16.7 E	5.40 W	8.0 W			

This table shows that at London, for example, the pointing of the needle was east of north in the middle of the sixteenth century, reaching a maximum of 11° or 11¼° about 1580. After that it began to diminish until about 1658, the year of Cromwell's death, when the needle stood truly north and south. The needle next began to point westward by an ever-increasing amount until about 1812, when it appeared to almost stand still for several years at a value of somewhat over 24°. Thereafter the westerly declination began to diminish until it is now about 16°. Consequently between 1580 and 1812, in an interval of 232 years, the compass direction at London changed from 11° east to 24° west, in all 35°. *The direction of a street a mile long, laid out in London in 1580 in the direction pointed out by the compass would be seven-tenths of a mile too far to the east at the north terminus according to the compass direction of 1812!*

For Paris and Rome similar changes to those at London are found. At Paris the maximum easterly declination of 9° 36' was reached near the year 1580, and the maximum westerly declination of 22° 36' in about 1809, the needle pointing due north in 1664. At Rome the declination of the needle reached its maximum amount east, 11° 36', in 1570, approximately, and its maximum amount west, 17° 06', in about 1810, coinciding with the true meridian in 1660. At Manila, Philippine Islands, the needle changed from 05' east in 1800 to 53' east in 1901, and at San Francisco, Cal., from 12° 36' east in 1780 to 16° 48' east at the present time. At Baltimore, between 1640 and the present time the needle bore west all the time and did not at any time point due

---

[a] This table and the accompanying subsequent remarks are extracted from the writer's "First report on magnetic work in Maryland," Maryland Geological Survey Report, Vol. I, 1897.

north or east of north as surveyors frequently assumed to be the case for this part of Maryland. The figures show that at Baltimore the compass needle pointed about 6° 06' west in 1670 and that in about 1802 it pointed the least amount west, namely, 39'; hence, in an interval of 132 years, the needle changed its direction by 5° 27'. *A street a mile long laid out in Baltimore in 1670 so as to run in the compass direction would have its north terminus 504 feet, or about one-tenth of a mile, too far to the west in 1802.* This is a fact especially interesting, because in some of the old towns of the thirteen original States, as for example in Maryland, the streets were laid out by the compass, or prominent public buildings, such as court-houses, were erected so that the front face would run parallel to a cardinal direction as given by the compass. Thus, while establishing a meridian line for the use of surveyors at Chestertown, the county seat of Kent County, Md., it was found that High street, the main street, ran very nearly magnetically northwest and southeast. Assuming that the street was originally laid out with the compass so as to run northwest and southeast, and knowing from the data at Baltimore and some other stations that the needle bore the same amount west in the early part of the eighteenth century that it does at present, the conclusion to be drawn was that the town of Chestertown was laid out in the early part of the eighteenth century. Upon looking up the records, the assumptions made and the conclusions drawn were verified. The town was laid out in 1702 and the streets were run with the compass northwest and southeast, and at right angles thereto. So, also, by determining the astronomical directions of the streets in the old town of Oxford, Md., which had been laid out by the compass in the first decade of the eighteenth century, an approximate knowledge of the magnetic declination at that time was ascertained.

The table likewise gives the change in the compass direction at some stations in the Southern Hemisphere. One fact at once noticeable from this table is, *that during a given interval of time the compass direction changes not only by different amounts in different parts of the Earth, but, likewise, the changes occur in some parts in opposite directions.* For example, compare the changes which have occurred between 1800 and 1890 at the various stations.

Place.	North end of compass needle veered between 1800 and 1890.		
London	6° 30'	to the east.	
Paris	6 58	"	"
Rome	6 29	"	"
Manila	0 45	"	"
San Francisco	3 02	"	"
Baltimore	4 14	"	west.
Rio de Janeiro	11 36	"	"
Ascension Island	9 00	"	"
St. Helena Island	6 42	"	"
Cape Town	3 48	"	"

The compass needle, accordingly, while swinging to the *eastward* at London between 1800 and the present time was swinging in the opposite direction, *westward*, at Baltimore during the same interval of time, the amount of swing not being the same at the two stations.

Another striking fact disclosed by looking over the figures for any one station, for example, Baltimore, is *that at the same station the change per year is not a constant quantity, as frequently assumed by the surveyor.* The annual change for this particular station

PRINCIPAL FACTS OF THE EARTH'S MAGNETISM. 43

may vary all the way from zero to four minutes. At the times of maximum or minimum values of the declination the annual change is practically zero for about five years on either side of these epochs. The annual change then begins to increase until about midway between the epochs of maximum and minimum values, for example, about 1730 or about 1870, when it reaches its maximum value of about four minutes; it then diminishes again.

The secular motion of the compass needle may be likened to the swinging of a pendulum. At the extreme positions of the pendulum, on either side of the position it would occupy if at rest, the velocity with which the bob moves in its orbital path vanishes. As the pendulum moves toward its mean position from the right, it does so at a constantly accelerating pace until it reaches the mean position midway between the two extreme positions. Here the velocity is a maximum, and as the pendulum swings past the mean position it begins to slacken its pace until reaching the extreme position on the left, when the velocity of motion again vanishes.

At no station has as yet a complete swing—for example, from right to left and back again from left to right—been observed. At some stations, however, a little over half a swing has been obtained. A comparison of the time interval between the two extreme positions, i. e., half a swing, at various stations shows another remarkable fact, *that the time intervals between the extreme positions of the needle are of different lengths in different parts of the Earth.* To illustrate: At London, Paris, and Rome the time interval between dates of extreme positions of the needle is about two hundred and thirty to two hundred and forty years, while for stations in the Eastern States of this country it is on the average about one hundred and fifty years.

Taking into consideration all the facts at present known with regard to the secular change, it is found that it is not possible to explain all those facts on the assumption that there is a secular change period common to all parts of the Earth of about three hundred to five hundred years in length. The indications are that for a common secular change period a much longer period is required. But if this is so, it means that the secular change is a far more complicated matter than generally supposed. Besides the main swing as described above, there are a number of minor swings whose periods are not as yet definitely known. These minor swings have the effect of slightly altering the annual change due to the main secular change.

Fig. 8 illustrates graphically the change in the magnetic declination for various points in the Northern Hemisphere, such stations having been selected as would be typical of the regions represented by them. It will be seen that the stations encircle the globe. This one diagram exhibits at a glance all the characteristic features of the secular change of the magnetic declination in the Northern Hemisphere as at present known. With the aid of Table II the meaning of the curves will be readily understood. Thus, for example, selecting the date 1800 and running the eye along the horizontal line marked 1800 until it intersects the London curve, and casting the eye upward from this point of intersection along the vertical line, it is found that the declination of the needle was a trifle over 24° west. For Paris the observations known up to the present time have been indicated by dots. It will be seen that the curve, which is due to Schott, represents the existing data satisfactorily. In the case of Fayal Island it will be noticed that prior to 1600 two curves, one in full and the other broken, are given; the broken curve represents a repetition of the same law which governed the secular change at this station between 1600 and present date, while the full curve has been drawn to harmonize with the observations back to the time of Columbus. It will be seen that there is a marked difference between the two curves for the date 1500. A similar state of things is revealed at Rome, the broken curve again representing the law from 1510 to present date, while the full curve represents the observations which can be obtained with the aid of the early "compass charts" of the fourteenth and fifteenth centuries. The departure between the broken curve and the full one amounts to about 17° for the year 1400. Similar indications exist at other stations of a change in the law of the secular change prior to 1600.

The special purpose of the diagram has been to show the mutual relationship

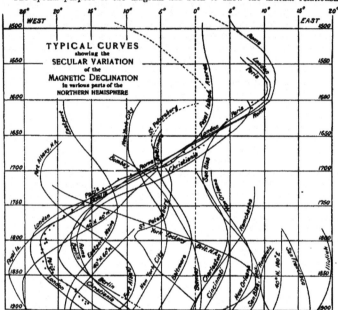

FIG. 8.—Comparison of the secular change curves of the magnetic declination at various stations in the Northern Hemisphere.

between the secular change curves over the Earth. Each station bears a somewhat different testimony of the phenomenon under consideration, and it is only by considering the collective evidence that one can hope to make headway and be enabled to say what probably transpired at any one station prior to the records or what is likely to occur at this station in the future. By following the curves systematically around the globe it is quite possible to construct a composite curve, with the aid of which a clearer conception of this most perplexing phenomenon can be obtained.

However, as already stated, *the laws actually governing the secular change can not be discovered by simply considering the changes in the magnetic declination alone.* One can hope to make progress only by studying the phenomenon in its *entirety;* that is to say, if a magnetized needle is taken and suspended at its center of gravity in such a way that it is free to turn in any direction whatsoever, to the left or to the right, up or down, then under the influence of the Earth's magnetism the *north* end of the needle, while still pointing approximately toward the north, also points *downward* and the *south* end *upward.* The actual direction assumed by the needle lies somewhere between a true vertical line and a true horizontal line, nearer to the former than to the latter in the latitudes under consideration. This is the direction in which the Earth's magnetic force acts. On the *compass needle* only the *horizontal* component of the force has an effect; as the *vertical* component is counteracted by adding an additional weight to the south arm of the needle, generally a bit of brass wire. The changes that are taking place in the

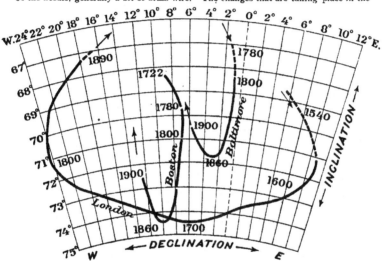

FIG. 9.—Curves showing secular change in magnetic declination and dip at London, Boston, and Baltimore.

true direction of the Earth's magnetic force and in its *magnitude* constitute the real facts to be studied.

It is an interesting problem to inquire: How does the north end of the freely suspended magnetic needle move with the lapse of time, if the motion is observed from the point of suspension of the needle? Does it move clockwise or anticlockwise? Would needles similarly suspended in all parts of the Earth move in the same direction? What is the nature of the curve described in space by the north end? These are some of the fascinating questions which can be asked from this point of view.

It has been found by the writer that over the greater portion of the Earth the north end of a freely suspended magnetic needle during the past two or three centuries has been moving in a *clockwise* direction. In the Pacific Ocean and along the western coast of the United States evidence exists of small irregularities in the general law of motion as explained above. Some of the stations in this region exhibit small *anticlockwise* motions. *No station has thus far been found where the reverse motion has prevailed for any such length of time, as has been the case with the direct motion.*

Fig. 9 exhibits the curves resulting in the manner described above for London, Boston, and Baltimore.

Fig. 10 has been constructed in a similar manner. The outside curve exhibits the changes in magnetic declination and dip encountered were one to make a complete

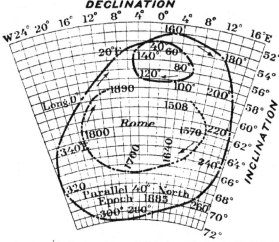

FIG. 10.—Comparison of curve showing change in magnetic declination and dip along parallel of latitude 40° N. in 1885 with curve showing secular change at Rome.

circuit of the Earth in an eastwardly direction along the parallel of latitude 40° north. The data have been scaled from Neumayer's isogonic and isoclinic charts for 1885, contained in his excellent atlas. Thus in zero longitude, counting from Greenwich, a freely suspended magnetic needle pointed in 1885 15½° west and its dip was 58°; in 20° east longitude, these quantities were respectively 8° west and 54°.7, etc. It will be noticed that the curve goes throughout—even for the loop described when crossing Asia—in the same direction as that of the hands of a watch, just as in the case of the secular motion curves shown in fig. 9 and the one of Rome given in the present figure. Rome is situated not far from latitude 40° north, its latitude being 41°.9 north. The general character of the two curves is seen to be very similar. It has been shown in other ways besides this one that many of the laws underlying the momentary distribution of the Earth's magnetism and the secular change are alike.

The circuit of the Earth in the above case was made to the eastward because the secular variation curves appear to develop themselves more and more as we go around the Earth eastwardly.[a]

---

[a] See Physical Review, Vol. II, pp. 455-465, and Vol. III, pp. 34-48.

PRINCIPAL FACTS OF THE EARTH'S MAGNETISM.

DIURNAL VARIATION.

In the year 1682, in the city of Louveau, Siam, it is related that Pater Guy Tachart, in the presence of the King, found that the magnetic declination on one day was 0° 16' west; on the following day, 0° 31'; on the third day, 0° 35'; on the fourth, 0° 38', and repeating the observations after the lapse of a few days the values found on three successive days were 0° 28', 0° 33', and 0° 21'. The observations were doubtless made on these various days at different times of the day, so that part of the differences in the results obtained are possibly to be ascribed to the next remarkable fact regarding the "constant inconstancies" of the Earth's magnetism, the so-called diurnal variation, by which the needle is made to change its direction, from hour to hour, throughout the day.

The credit of the discovery of the diurnal variation must properly be given to Graham, a London mechanician and clock maker, who from many hundred observations of the declination of the magnetic needle at various times of the day made in 1722 a definite announcement of the existence of this variation.[a] Graham's discovery was later verified and amplified by Prof. Andr. Celsius in Upsala, who had a compass made expressly for this purpose by the instrument maker, Sisson, of London, under Graham's supervision, and by a host of other investigators.

TABLE III.—*The diurnal variation of the magnetic declination at Baldwin, Kans., for each month of the year 1901.*

Hour	Jan.	Feb.	Mar.	Apr.	May	June	July	Aug.	Sept.	Oct.	Nov.	Dec.
1 a. m.	−0.4	−0.2	+0.3	+0.5	+0.3	+0.1	+0.3	+0.1	0.0	−0.1	−0.1	−0.1
2	−0.4	−0.2	+0.2	+0.6	+0.5	+0.2	+0.3	+0.2	0.0	−0.1	−0.2	−0.2
3	−0.2	0.0	+0.2	+0.5	+0.6	+0.4	+0.4	+0.1	+0.3	+0.2	−0.1	−0.2
4	−0.1	+0.2	+0.4	+0.8	+0.8	+0.8	+0.8	+0.3	+0.8	+0.2	0.0	0.0
5	−0.5	+0.4	+0.4	+0.9	+1.4	+1.4	+1.4	+1.2	+1.1	+0.4	+0.2	−0.2
6	−0.3	+0.3	+0.8	+1.6	+2.2	+2.4	+2.4	+2.7	+2.3	+0.8	+0.3	−0.2
7	0.0	+0.9	+1.9	+2.7	+3.1	+3.5	+3.6	+4.2	+4.0	+2.0	+1.2	+0.3
8	+0.6	+1.5	+2.5	+3.5	+3.3	+3.7	+4.0	+4.4	+3.6	+2.6	+2.0	+0.4
9	+1.5	+2.2	+2.7	+2.8	+2.8	+3.2	+3.4	+3.3	+2.4	+2.4	+2.0	+1.2
10	+2.0	+1.3	+1.8	+1.1	+1.0	+1.0	+0.9	+0.3	0.0	+1.0	+1.0	+1.4
11	+1.1	+0.1	−0.2	−0.8	−1.1	−1.2	−1.7	−2.2	−2.2	−0.8	−0.6	+0.6
Noon	−0.4	−1.2	−2.0	−2.1	−2.4	−2.6	−3.2	−3.7	3.6	−2.2	−1.7	−0.7
1 p.m.	−1.3	−1.8	−3.0	−3.1	−3.4	−3.5	−4.0	−4.3	−3.8	−2.5	−2.1	−1.5
2	−1.6	−2.0	−3.0	−3.5	−3.7	−3.7	−4.1	−4.1	−3.0	−2.1	−1.8	−1.6
3	−1.4	−1.6	−2.5	−3.0	−2.8	−2.9	−3.1	−3.0	−1.5	−1.4	−1.1	−1.3
4	−1.1	−0.9	−1.5	−2.0	−1.8	−1.8	−1.9	−1.3	−0.4	−0.7	−0.5	−0.7
5	−0.3	−0.3	−0.5	−1.0	−0.8	−0.6	−0.7	0.0	+0.1	−0.4	−0.1	+0.1
6	+0.2	0.0	−0.1	−0.3	−0.1	−0.1	+0.2	+0.5	−0.2	−0.2	+0.1	+0.1
7	+0.5	+0.2	−0.1	0.0	−0.2	0.0	+0.2	+0.2	−0.1	+0.2	+0.4	+0.5
8	+0.6	+0.4	+0.1	0.0	−0.1	0.0	+0.2	+0.1	+0.2	+0.2	+0.4	+0.6
9	+0.6	+0.5	+0.4	0.0	−0.1	0.0	+0.2	+0.2	+0.4	+0.3	+0.4	+0.6
10	+0.5	+0.2	+0.5	+0.2	0.0	−0.1	+0.2	+0.3	0.0	+0.2	+0.1	+0.4
11	+0.4	+0.2	+0.4	+0.2	+0.1	0.0	+0.2	+0.2	0.0	+0.2	+0.1	+0.2
Mid't	+0.1	0.0	+0.3	+0.6	+0.2	−0.1	+0.1	+0.3	−0.3	0.0	0.0	−0.1
Range	3.6	4.2	5.7	7.0	7.0	7.4	8.1	8.7	7.8	5.1	4.1	3.0

[a] See "Philosophical Transactions," London, 1724.

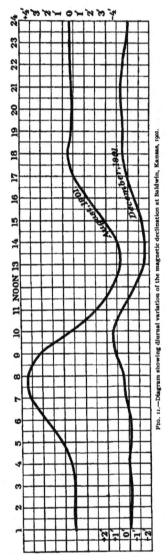

Fig. 11.—Diagram showing diurnal variation of the magnetic declination at Baldwin, Kansas, 1901.

Table III shows how the compass needle changed its direction from hour to hour (local mean time) for each month of the year 1901 at Baldwin, Kans., where the Coast and Geodetic Survey has a magnetic observatory in which are mounted delicate instruments registering continuously, day and night, automatically, by photographic means, the minutest variations in the Earth's magnetism.

At that place the magnetic needle points about 8°.4 east of north. A plus sign in the table means a deflection of the needle toward the east of the average direction for the entire day (twenty-four hours), and a minus sign a deflection toward the west. Thus in August, for example, at 8 a. m. the average easterly pointing of the needle was increased by 4'.4; it then began to diminish until the average value was reached a little after 10 a. m., indicated by the change of sign of the tabular quantities; after passing this point it still continued to diminish until reaching its lowest value at about 1 p. m., when the easterly declination had its least value, being 4'.3 less than its average value, or about 9' less than its maximum value in the morning. Next it increased until again reaching its average value about 5 p. m., after which it remained nearly stationary, except for minute fluctuations throughout the night, until about sunrise, when it rapidly began to rise to its maximum value.

Examining the figures for a winter month, e. g., December, it will be seen that the fluctuations are not so large as during the summer; where before the difference between maximum and minimum was about 9', it is now one-third of this amount, viz, 3'. On the diagram, Fig. 11, the diurnal variation of the magnetic declination for the two months, August and December, has been graphically represented.

Two lines, each a mile long, one run in the direction indicated by the compass early in the morning and the other early in the afternoon, both starting at the same point, diverge at their extremities in midsummer by 10–15 feet, the morning line being to the east of the afternoon one; in midwinter the divergence would be about one-third of this amount. It will thus be seen

PRINCIPAL FACTS OF THE EARTH'S MAGNETISM.   49

that the diurnal variation is of sufficient importance to be taken into account in accurate land surveys.

In Germany mine surveying has become such an art that some of the principal mines maintain small magnetic observatories, where the declination is recorded continuously throughout the day by photographic means. The mine surveyor then uses the value of the declination to the nearest minute prevailing at the time of day when he is running his line.

Where the needle points west of north, the times of maximum and minimum value of the magnetic declination will be reversed from what they are at Baldwin, the minimum occurring in the morning and the maximum in the afternoon. Of the two lines a mile long, considered above, the morning line will, however, again be east of the afternoon line.

The times when the declination reaches its extreme values, or when it reaches its average value, as is evident from Table III, are subject to fluctuations in the course of the year, being retarded during the months when the sun is south of the equator. These changes, which undergo a complete cycle in the course of one year, likewise manifest themselves in the magnitude of the diurnal range, approximately shown by the figures given in the bottom line of the table.

The approximate local mean time when the average declination is reached, in the United States is, on the average for the year, at about 10:30 a. m., and again about about 6 p. m. (See next table.)

The following comparative table, No. IV, of the diurnal variation was prepared by Schott[a] in order "to exhibit the changes which the total solar-diurnal variation undergoes with a change of geographical position within the region of North America." The series of observations which he admitted "extend over one or more years, and in no instance have any so-called disturbances been excluded." "The year or years of each series is added to admit of a correction for position in the sun-spot period."

The particulars for each station are as follows:

Name	Latitude	Longitude W. of Gr.	Magnetic Dip	Diurnal Range of Declination	Extent of series
	° ′	° ′	° ′	′	
Key West, Fla.	24 33.1	81 48.5	54 32	4.7	Mar., 1860, to Mar., 1866, exclusive
Los Angeles, Cal.	34 03.0	118 15.4	59 30	5.8	Oct., 1882, to Oct., 1889, exclusive
Washington, D. C.	38 53.6	77 00.6	71 19	7.5	July, 1840, to June, 1842, inclusive
Philadelphia, Pa.	39 58.4	75 10.2	71 58	7.8	Jan., 1840, to June, 1845, inclusive
Madison, Wis.	43 04.5	89 24.2	73 56	6.7	Mar., 1877, to Mar., 1878, exclusive
Toronto, Canada	43 39.4	79 23.5	75 15	8.8	July, 1842, to June, 1848, inclusive
Sitka, Alaska	57 02.9	135 19.7	75 55	10.6	Irregular series, 1848 to 1862
Uglaamie, Point Barrow	71 17.7	156 39.8	81 24	40.1	Sept., 1882, to Aug., 1883, inclusive
Plover Point, Point Barrow	71 21.4	156 16.1	81 36	38.6	17 months, 1852–1854
Fort Rae, Great Slave Lake	62 38.9	115 13.8	82 54	41.4	Oct., 1882, to Sept., 1883, inclusive
Kingua Fjord, Cumberland Sound	66 35.7	67 19.2	83 51	43.7	Do.
Fort Conger, Grinnell Land	81 44.0	64 43.8	85 01	98.8	Sept., 1881, to Aug., 1882, inclusive

[a] See Appendix No. 9, Coast and Geodetic Survey Report for 1890, pp. 261–264.

TABLE IV.—*Total solar-diurnal variation of the magnetic declination, on the yearly average, at prominent places in North America.*

[A + sign indicates a deflection of the north-seeking end of the magnet toward the *east*, a − sign the contrary direction.]

Local mean time.	1. Key West, Fla.	2. Los Angeles, Cal.	3. Washington, D.C.	4. Philadelphia, Pa.	5. Madison, Wis.	6. Toronto, Canada	7. Sitka, Alaska	8. Ugiaamie, Point Barrow	9. Plover Point, Point Barrow	10. Fort Rae, Great Slave Lake	11. Kingua Fjord, Cumberland Sound	12. Fort Conger, Grinnell Land	Average values, stations 1 to 6, inclusive
	′	′	′	′	′	′	′	′	′	′	′	′	′
1 a. m.	+0.0	+0.0	+0.7	+0.6	+0.1	+0.6	+0.2	−12.8	− 8.0	−11.0	+11.7	+43.2	+0.35
2 a. m.	−0.0	+0.1	+0.7	+0.5	0.0	+0.5	+1.0	− 4.9	− 1.9	− 6.6	+15.8	+45.1	+0.05
3 a. m.	+0.1	+0.2	+0.9	+0.6	+0.2	+0.8	+1.4	+ 3.3	+ 3.6	+ 0.8	+18.0	+41.2	+0.07
4 a. m.	+0.2	+0.3	+1.2	+1.0	+0.5	+1.2	+2.0	+ 6.2	+10.9	+ 7.4	+19.1	+25.7	+0.75
5 a. m.	+0.4	+0.6	+1.7	+1.5	+1.0	+1.8	+2.9	+14.3	+16.6	+13.6	+19.3	+31.6	+1.19
6 a. m.	+1.0	+1.3	+2.1	+2.1	+1.4	+2.7	+4.2	+21.6	+19.3	+21.0	+20.1	−19.7	+1.79
7 a. m.	+2.1	+2.4	+2.8	+3.3	+2.6	+3.5	+5.3	+26.1	+27.1	+26.2	+19.9	+26.6	+2.80
8 a. m.	+2.5	+3.1	+3.2	+3.5	+3.2	+3.8	+6.0	−26.7	−27.0	−29.4	+17.4	−18.7	+3.24
9 a. m.	+2.2	+2.6	+2.3	+2.8	+3.0	+3.0	+5.3	−26.1	−19.9	+25.5	+10.8	+ 1.2	+2.67
10 a. m.	+1.1	+1.1	+0.9	+0.8	+1.7	−0.8	+3.0	+ 9.9	+ 9.3	+16.8	+ 3.7	−12.7	+1.09
11 a. m.	−0.2	−0.8	−1.3	−1.6	−0.7	−2.0	+0.6	+ 1.4	− 0.4	+ 8.0	+ 1.3	−21.4	−1.08
Noon	−1.4	−2.2	−3.2	−3.4	−2.5	−4.2	−2.1	− 5.9	− 8.2	− 0.9	− 9.0	−40.7	−2.80
1 p. m.	−2.1	−2.7	−4.3	−4.3	−3.5	−5.0	−3.2	− 7.3	−10.7	− 4.0	−15.1	−45.6	−3.63
2 p. m.	−2.2	−2.6	−4.3	−4.1	−3.5	−4.8	−4.2	− 7.7	− 9.8	− 8.1	−21.2	−49.2	−3.56
3 p. m.	−1.9	−2.0	−3.5	−3.1	−2.6	−3.8	−4.6	− 7.3	− 9.9	−10.6	−20.4	−45.8	−2.80
4 p. m.	−1.3	−1.1	−2.5	−2.2	−1.6	−2.5	−4.6	− 9.1	− 9.8	−11.3	−20.6	−53.7	−1.85
5 p. m.	−0.8	−0.5	−1.5	−1.0	−0.7	−1.3	−3.8	− 9.9	−10.2	−12.1	−23.6	−23.7	−0.95
6 p. m.	−0.4	−0.2	−0.8	−0.4	−0.2	−0.3	−3.2	− 9.9	− 9.7	−12.9	−19.4	−17.3	−0.36
7 p. m.	−0.2	−0.0	0.0	+0.0	+0.2	+0.2	−2.4	− 8.4	− 8.4	−12.5	−16.1	−27.2	+0.05
8 p. m.	+0.1	+0.1	+0.6	+0.6	+0.2	+0.7	−1.4	− 6.0	− 9.0	−11.0	−15.5	− 3.5	+0.44
9 p. m.	+0.2	+0.1	+1.0	+0.6	+0.6	+1.2	−0.8	− 8.1	− 7.5	−12.0	− 8.8	+ 3.5	+0.64
10 p. m.	+0.2	+0.1	+1.1	+1.2	+0.7	+1.3	−0.4	−10.9	− 7.9	−11.9	− 0.6	+22.4	+0.79
11 p. m.	+0.2	+0.1	+1.1	+0.7	+0.2	+1.2	−0.6	− 9.1	−11.5	−11.9	+ 3.9	+30.0	+0.60
Midnight	+0.1	+0.0	+1.0	+0.6	+0.1	+0.8	−0.6	−13.4	−10.8	−12.0	+ 9.2	+32.6	+0.45
Range	4.7	5.8	7.5	7.8	6.7	8.8	10.6	40.1	38.6	41.4	43.7	98.8	6.9

Schott's deductions from this table are:

"A perusal of the tabular values for the localities marked 1 to 6, and which represent all that part of the United States and Canada which lies south of the forty-ninth parallel, shows a very close accord of the diurnal variation, having an average maximum easterly deflection of 3′.2 at about 7ʰ.9 in the morning and an average maximum westerly deflection of 3′.6 at about 1ʰ.4 in the afternoon, although the dip varies 20¾° between these geographical limits. At Sitka the range reaches already 10′.6 and beyond, with a dip of 80° and more, the diurnal range rapidly rises, attaining 1° 40′ nearly at Fort Conger. At the higher (magnetic) latitude stations there is a tendency to shift the morning extreme to an *earlier* hour and the afternoon opposite extreme to a later hour than the corresponding times just cited. A remarkable feature in the diurnal variation (yearly average) is the close correspondence in the local times when the needle passes the average magnetic meridian (tabular values passing from + to − sign); these times are:

	h	m
For Key West	10	51
Los Angeles	10	35
Washington	10	25
Philadelphia	10	20
Madison	10	43
Toronto	10	17
Average	10	32

"This time is subject to an annual inequality which at Los Angeles in the summer months displaces it to about $10^h \ 00^m$, and in the winter months to about $11^h \ 30^m$."

The diurnal range of the magnetic declination as is seen in Table IV, increases with an approach to the magnetic pole and decreases toward the magnetic equator. If d represents the diurnal range, I, the dip and ϕ, the "magnetic" latitude as found from the formula $\tan \phi = \frac{1}{2} \tan I$, then the following formula:

$$d = 2'.58 \sec^2 \phi$$

will give a fair representation of the law according to which the diurnal range varies with magnetic latitude or dip.

The diurnal range increases with an approach toward the magnetic pole because the horizontal component of the magnetic force, which holds the compass needle in place, diminishes with a movement in this direction, whereas the deflecting forces which cause the diurnal variation increase, and thus their effect increases with increase of magnetic latitude. The only horizontal force acting on the compass needle at the magnetic pole is that due to the diurnal variation, and to magnetic perturbations, so that, if the needle were suspended with sufficient delicacy it might pass back and forth through all points of the compass in the course of the day.

The average value, for the year, of the diurnal range is subject to a mysterious fluctuation, being greater in years of maximum frequency of sun spots, and less in times of minimum frequency or minimum solar activity as exhibited by sun spots. The next table, V, shows this. The numbers in column R, due to Wolf, represent the variation for the years given in the sun-spot frequency. Thus in the year 1843, a minimum sun-spot year, it is found that the range of declination at Philadelphia reached its smallest value. The period 1883–1884 was a maximum sun-spot year, and it is seen that the range at Los Angeles reached its maximum value during this time.

TABLE V.—*Showing how the diurnal range of the magnetic declination varies during the sun-spot period—(about 11 years).*

	Philadelphia			Los Angeles		
Year	d Diurnal range of declination	R Relative sun-spot frequency numbers		Year (Oct. to Oct.)	d Diurnal range of declination	R Relative sun-spot frequency numbers
1840	9.1	61.8		1882–83	6.5	60.7
1841	8.1	38.5		1883–84	7.1	68.2
1842	7.8	23.0		1884–85	6.9	53.7
1843	7.5	13.1		1885–86	5.8	32.4
1844	7.5	19.3		1886–87	5.4	14.3
1845	8.5	38.3		1887–88	5.4	7.3
				1888–89	5.1	7.4

According to the researches of two Russians, Leyst and Passalskij, the diurnal variation is different over locally disturbed areas, e. g., in regions of iron mines, from what it would be if the disturbances did not exist. Hence in such regions, the continuous records of distant magnetic observatories can not be utilized for referring the magnetic elements to the mean value for the day, or to some other period of time, but, special observations for this purpose must be made in the disturbed locality. Whether the secular change is likewise different over locally disturbed regions from what it would be if the local disturbance were not present, is not yet known.

Just as the declination suffers change from hour to hour throughout the day, so likewise are the other elements of the Earth's magnetism, the dip and the intensity, affected.

The diurnal variation, as has been shown, progresses according to the hours of local mean time, or, in other words, is connected in some manner with the Earth's rotation whereby different parts of its surface are exposed to the action of the Sun's rays, and it may be presumed, therefore, that the Sun plays a prominent part in causing the daily variation in the Earth's magnetic state. The precise manner in which the Sun brings about this variation has not yet been satisfactorily explained in spite of the researches of the most eminent investigators. The most commonly accepted opinion is that the diurnal variation is due to a peculiar system of electric currents in the upper regions of the atmosphere, the precise way in which their existence is brought about not being, however, as yet clear.[a]

The diurnal variation furnishes the first evidence that the Earth's magnetism is in close touch with outside influences and responds in a most mysterious and sympathetic manner with changes ever going on in the upper regions. The facts related in the following pages give further evidence on this subject.

ANNUAL VARIATION.

If the monthly values of the magnetic declination be corrected for the secular change in the course of the year, they exhibit a slight variation, having the year as the period, known as *the annual variation of the magnetic declination*. This is not to be confounded with the *annual change* of the declination, which means the change in one year due to the secular variation. The latter is a *progressive* change, so that the needle at the end of the year does not point the same way that it did at the beginning, while the annual variation is a *cyclical* change, that is, as far as the annual variation is concerned, the needle returns to the same position virtually at the end of the year that it had at the beginning. The next table shows how minute a quantity this annual variation is, and that it can be neglected for all practical purposes.

---

[a] The reader who is interested in the latest theoretical developments might be referred to Schuster's paper in Phil. Trans. R. S., Part A, 1889; von Bezold's papers, Berlin Academy of Sciences, 1897, and Nippoldt's papers, Terrestrial Magnetism, Vol. VII. A summary of Schuster's and von Bezold's researches will be found in Gray's Magnetism and Electricity, Vol. I, 1898.

TABLE VI.—*Annual variation of the magnetic declination at several places in the northern magnetic hemisphere.*[a]

[A + sign denotes a deflection of the north end of the magnet to the *eastward*, a — sign, the contrary direction.]

Month	Los Angeles, Cal. 1882–1889	Key West, Fla. 1862–1865	Washington, D. C. 1840–1842, 1867–1866	Philadelphia, Pa. 1840–1845	Toronto, Canada 1845–1851, 1856–1864, 1865–1871	Dublin, Ireland 1841–1850	Kew, England 1858–1862
January	+0.6	—0.6	+0.6	—0.5	0.0	+0.4	0.0
February	+0.2	—0.6	—0.3	—0.4	+0.2	+1.6	—0.6
March	—0.4	—0.1	+0.2	+0.1	+0.1	+1.7	—0.5
April	—0.4	+0.3	—0.1	+0.1	0.0	+1.9	0.0
May	—0.4	+0.3	—0.4	—0.2	+0.3	+1.3	+0.7
June	—0.4	+0.2	—0.1	+0.6	+0.5	0.0	+0.8
July	—0.4	+0.3	+0.2	+1.0	+0.4	—1.2	+1.2
August	—0.1	+0.8	+0.7	+0.9	0.0	—2.2	+0.3
September	+0.3	+0.7	—0.4	0.0	—0.4	—2.1	—0.2
October	+0.4	—0.5	—0.2	+0.2	—0.6	—1.4	—0.8
November	+0.5	—0.5	—0.2	—0.9	—0.4	—0.3	—0.6
December	+0.6	—0.3	—0.3	—0.7	—0.1	+0.2	—0.7

It is seen that the total range of the annual variation is a very small quantity, about 1' for the North American stations. The character of the variation appears to be different for each station. This may possibly be because the tabular results do not refer in each case to the same interval of time, and because they were not deduced by one common method.

According to the recent investigations of Dr. Schwalbe, the forces which bring about this variation are situated outside the earth.

MINOR PERIODIC FLUCTUATIONS.

Chief among these may be mentioned the variation depending upon the position of the Moon with reference to the Sun and the Earth. The range, or difference between the extreme values, of this variation is so minute that it has required many years of continuous and carefully made observations to detect it.

MAGNETIC STORMS.

Generally speaking these may occur at any time and are frequently accompanied by auroral displays. Such storms may at times have a very wide circle of action and occur practically simultaneously over the whole area. Thus on December 3, 1896, while the writer was on his way to Salisbury, Md., to make magnetic observations, he saw a most brilliant aurora, and the next day while making magnetic observations the behavior of the needle indicated that a magnetic storm was prevailing. This storm it was afterwards ascertained occurred at foreign observatories practically simultaneously with its occurrence in Maryland.

The fluctuations caused by these spasmodic variations in the Earth's magnetism may in the United States occasionally amount to as much as 10–20' and even more.

---

[a] From Coast and Geodetic Survey Report for 1890, p. 249. The matter contained in Tables IV and V was taken from the same source.

Thus, on October 12, 1896, the writer made observations at Oakland, Md., at various times during the day. The diurnal variation on that day was completely reversed, the maximum value of west declination occurring in the morning instead of in the afternoon, and the minimum value in the afternoon instead of the morning. The observation in the morning required a correction of $-16'$.

Small, spasmodic fluctuations occur frequently; in fact, scarcely a day passes without them.

It is due largely to these irregular disturbances, the coming of which can not be predicted, that it is not possible to give by a general system of rules accurate reductions of an observed declination to the mean value for the day.

The duration of the irregular fluctuations may be but an instant, a few hours, or several days. They generally reveal their presence by a sudden and marked departure of the needle from its true normal position. While these fluctuations make their appearance apparently at random, nevertheless when they are treated statistically it is found that they exhibit well-marked periodicities in their occurrences. They are more frequent and more violent in the years of maximum solar activity, as indicated by sun spots, and less frequent and less violent in years of minimum activity. In November, 1882, near the period of maximum sun spots, a magnetic storm occurred which caused the magnetic needle at Los Angeles, Cal., to move over $1\frac{1}{2}°$ out of its normal position. There was at the time a brilliant auroral display. This storm occurred over the entire Earth, at Los Angeles, Toronto, London, St. Petersburg, Bombay, Hongkong, and Melbourne, and began at practically the same instant of absolute time. Then again they appear subject to periodic variations, such as the daily and the annual. They apparently occur more frequently toward evening and less frequently toward noon; also more frequently in equinoctial months and less frequently in solstitial months. Perhaps a good idea of the frequency and magnitude of the irregular disturbances is obtained from Schott's table,[a] based on the observations made every two hours at Philadelphia, under Bache, between the six years 1840 to 1845.

Deviations from normal direction.	Number of disturbances.
3. 6 to 10. 8	2189
10. 8 to 18. 1	147
18. 1 to 25. 3	18
25. 3 to 32. 6	3
Beyond,	0

It should be recalled that the period of minimum sun-spot activity occurred in the midst of this series; otherwise the disturbances would have been more frequent and greater. Schott cites the following maximum deflections:

	°	′
At Key West, between 1860 and 1866	0	21.4
At Madison, Wis., on May 28, 1877	0	24
At Madison, Wis., on October 12, 1877	0	48
At Lady Franklin Bay, during great storm in November, 1882, Greely noted a deflection of	20	28

---

[a] Coast and Geodetic Survey Report for 1888, App. 7.

G. R. Putnam, of the Coast and Geodetic Survey, cites a change of over $3°$ in twenty minutes at Niantilik on September 18, 1896. "At $7^h\ 35^m$ a. m. local mean time, the needle pointed $60°\ 35'$ west of north, while at $7^h\ 55^m$, it pointed $63°\ 50'$ west of north, and the total range for the day was over $4\frac{1}{2}°$. On this date there was an unusual magnetic disturbance, the extreme range in declination at Washington being $38'$ for the entire day, and $19'$ for the portion of the day corresponding to the interval during which observations were made at Niantilik. It will be noted that the range in declination was nearly fifteen times as great as at Washington during the same interval." The geographical position of Niantilik is $64°\ 53.'5$ north and $66°\ 19.'5$ west of Greenwich, and the dip on September 18, 1896, was $83°\ 54.'8$.

Some other interesting cases of magnetic storms will be given in the section on "Magnetic Observatories."

The cause of these remarkable phenomena of the Earth's magnetism whereby the whole magnetic system of the Earth is deranged at a moment's notice is shrouded in mystery. There are clearly three kinds of magnetic storms: (1) Cosmic ones, due to changes occurring in the regions above; (2) telluric ones, resulting from changes within the interior of the Earth, and (3) regional or local ones, resulting from changes within or external to the Earth's crust, whose field of action is limited to a restricted region of the Earth and the center or focus of which, while sometimes stationary, generally travels from place to place.

The principal phases of a storm of the first kind occur simultaneously over the Earth, within one or two minutes of time. Doubtless if arrangements could be made to time these principal phases at places over the entire Earth with greater accuracy than the ordinary photo-magnetic records will admit of, the correspondence in time would be within a few seconds. During the prevalence of these magnetic storms strongly marked variations in the electric currents within the Earth's crust manifest themselves along with the variations of the magnetic needle. Lord Kelvin has calculated the amount of energy required to produce the magnetic storm of June 25, 1885, if it were to be referred to direct action of the sun. Quoting from Gray's Magnetism and Electricity:

"The horizontal force at the following eleven places: St. Petersburg, Stonyhurst, Wilhelmshaven, Utrecht, Kew, Vienna, Lisbon, San Fernando, Colaba, Batavia, and Melbourne, increased considerably from 2 to 2.10 p. m., and fell from 2:10 to 3 p. m., with irregular changes in the interval.

"The mean value at all these places was .0005 above par at 2:10 and .005 below par at 3 p. m. The changes as shown by the photographic records were simultaneous at the different places. Assuming these electrical oscillations of the Sun, Lord Kelvin estimates that the electrical activity of the Sun during the storm, which lasted about eight hours, must have been about 160×10^{14} horsepower, or about 12×10^{25} ergs per second; that is, about 364 times the activity of the total solar radiation, which is estimated at about 3×10^{33} ergs per second. The electrical energy thus given out by the Sun in such a storm would supply, if transformed to the electrical vibrations of shorter period concerned in its ordinary radiation, the whole light and heat radiated during a period of four months. This, as Lord Kelvin remarks, is conclusive against the hypothesis that these violent magnetic disturbances are due to direct action of the Sun."

The probability is that a solar ray endowed with greater or less energy than ordinarily and of the necessary kind acted as the "trigger to the gun" to set off mighty electric forces whose presence in the upper regions is becoming more and more manifest every day.

A magnetic storm of the second category is associated with changes within the Earth, cataclysms, earthquakes, volcanic outbreaks, etc. The phases may occur simultaneously over very large portions of the Earth, or progress from place to place according to a certain rate. Remarkable coincident effects observed during the May eruption in Martinique will be found further on. Hansteen declared "that the variations of the magnetic needle are a mute language revealing to us the changes perpetually going on in the interior of the Earth." Another great student of nature, Clerk Maxwell says: "The never-resting heart of the Earth traces in telegraphic symbols the record of its pulsations, and also the slow but mighty working of the changes which warn us not to suppose that the inner history of our planet is ended."

Magnetic disturbances of the third kind, as stated, take place over a limited area, and are associated with phenomena occurring within the Earth, as enumerated in the previous paragraph, or with phenomena in the upper regions. In the case of these storms the passing of the principal phases from place to place may take a measurable amount of time. Storms of the first and second kind may bring about storms of the third kind as secondary phenomena.

Dr. Schmidt made a mathematical analysis of various magnetic storms, and in particular of the one which occurred on February 28, 1896, and whose course was followed one hour, from 6 to 7 p. m., Greenwich time, at the suggestion of Professor Eschenhagen, simultaneously by 15 observatories distributed over the Earth. His investigations clearly showed that the disturbance vectors at times converged to a point, at other times radiated from a point, and in times of magnetic calms (comparatively speaking) the vectors at the various stations were almost parallel to each other, as though pointing to a distant force center; furthermore, that the points of convergence in general moved progressively forward with a velocity of about 1 kilometer in a second, and also that they were at times nearly stationary. In view of the fact that the cause of the diurnal variation of the Earth's magnetism must apparently be referred to electric currents in the upper regions of the atmosphere, Dr. Schmidt believes that the immediate cause of the magnetic storms is to be referred to electric whirls or vortices which separate themselves from the general electric field in the atmosphere just as do the cyclones and anticyclones known to meteorologists. Taking also into consideration the vertical disturbing components and applying Ampere's rule to the current systems revealed by the disturbing forces, it follows that for the greater part of our observed magnetic storms the causes come from the outside of the Earth's crust.

MAGNETIC OBSERVATORIES.

These institutions are designed especially to secure a record of the changes ever going on in the magnetic condition of the Earth. It was recognized at an early date that the problems of terrestrial magnetism, like those of meteorology, have a world-wide interest and bearing, and so require for their successful and complete solution the united and harmonious efforts of all nations.

FIG. 12.—COAST AND GEODETIC SURVEY MAGNETIC OBSERVATORY AT CHELTENHAM, MARYLAND.

Under the powerful initiative of von Humboldt, Gauss, Herschell, Kupffer, and Sabine, a number of institutions were accordingly established in the fourth decade of the last century in different parts of the Earth, whose special purpose it was to record the ever-occurring magnetic variations. To cooperate with these foreign observatories a magnetic observatory—due to the enthusiasm of Professor Bache—was founded in this country at Girard College, Philadelphia. The results from these observatories amply showed the wisdom of international cooperation. At the present time[a] a special effort at a systematic study of the magnetic variations, according to a uniform plan, has again been originated, this time in cooperation with the various Arctic and Antarctic expeditions.

The Coast and Geodetic Survey has at present[a] four magnetic observatories taking part in this international work, viz, at Cheltenham, Maryland, 17 miles southeast of Washington; at Baldwin, Kansas, 17 miles south of Lawrence; at Sitka, Alaska, and in the Hawaiian Islands, at a site about 14 miles west of Honolulu. The first named, the Cheltenham Observatory, is one of the most complete and elaborately constructed magnetic observatories in existence, and its scope of work will include, besides the observation of magnetic phenomena, also seismic ones, and such as are related to atmospheric and to telluric electricity.

The illustration, Fig. 12, gives a view of the Cheltenham Magnetic Observatory, the larger building being the so-called "Variation Observatory," in which are mounted the self-registering photo-magnetic instruments, and the smaller building containing the office in the middle, flanked by two wings in which the absolute magnetic observations are made. The Variation Observatory consists of two rooms, each 16 by 19 feet; in the north room is mounted a magnetograph of the Eschenhagen pattern, and in the south room has been installed the Adie magnetograph, adapted for photographic registration and for eye readings, formerly at Los Angeles (1882–1889) and at San Antonio (1890–1895).

As the variations in the intensity of the magnetic force recorded on magnetic instruments are partly due to the changes in the magnetic moment of the suspended magnets due to temperature changes, it is necessary to either provide some means for determining these artificial changes and make corrections, or to institute the necessary arrangements for preserving a constant temperature in the observing room.

In the case of the Cheltenham Observatory, the attempt has been made to secure in an above-ground structure freedom from moisture and a uniformity of temperature within certain practical limits without employing any other means than that derived from the insulation of the specially constructed walls of the variation observatory building. In addition, thermographs register continuously any remaining temperature fluctuations inside the magnetograph rooms, with the aid of which any necessary reductions of the magnetic intensity variations to a selected standard temperature can be made. The drawing of the plans and the erection of the observatory were intrusted to J. A. Fleming, of the Coast and Geodetic Survey, and the results obtained thus far show that his method of construction was a successful one.

The wall insulation of the variation observatory is as follows: Beginning at outside of building, pine weatherboarding, 8-ply building paper, 1-inch pine sheathing, 8-inch air shaft, 1-inch pine sheathing, 8-ply paper, 3 feet pine sawdust, 8-ply paper, ⅞-inch pine ceiling, 3 feet 2 inches air space of passageway, ⅞-inch pine ceiling, 8-ply paper, 1 foot pine sawdust, 8-ply paper, ⅞-inch pine ceiling; slat ventilators or louvre

---

[a] 1902.

windows, so arranged and provided with closely fitting shutters that during the winter the 8-inch air shaft referred to can be made practically air-tight, while during summer when opened these tend to admit of the passage and circulation of a cooling draft around building. The insulation beginning at the roof and going down is: Gravel and asphalt pitch roof, 1-inch pine sheathing, 3 feet 8 inches air space communicating with 8-inch air shaft around building and provided with six louvre windows with close-fitting shutters as on those at bottom of air shaft, 1 inch rough pine floor, 3-foot filling of pine sawdust, 8-ply paper, ⅞-inch pine ceiling, 3-foot air space above rooms, 1-inch rough pine floor, 1 foot 6 inches pine sawdust, 8-ply paper, ⅞-inch ceiling. Insulation from bottom of foundation is 2 feet 8 inches of earth, 6-inch to 8-inch layer of screened gravel, about 3 feet pine sawdust, 1-inch pine under floor, 8-ply paper, ⅞-inch pine tongue-and-groove floor.

The greatest danger to fulfillment of the above results lay in the necessity of providing openings through walls for ventilation of rooms and for means of ingress and egress. Four shafts, each 5 by 10 inches and about 16 feet long, furnish air supply to passageway through wooden floor grates. These are provided with heavy rabbeted shutters made to fit very closely and fitted with refrigerator fasteners, so that they may be made air-tight. They are also provided at inlet with copper-wire screens of double thickness to break force of a wind blowing toward opening and to keep out such vermin as field mice. Ventilation of passageway is effected by four shafts opening into air space below roof, each 6 by 10 inches and about 16 feet long, provided with close-fitting slides. Ventilation of air space below roof is effected by three 14-inch copper "Star" ventilators. By the judicious use of these air-supplies and ventilators the danger of direct conduction of temperature changes through shafts can be entirely eliminated. Ventilation of magnetograph rooms from and into passageway is effected in each room by four 3-inch square vertical shafts in sawdust packing having inlet or outlet just below ceiling or above baseboard, according to arrangement of four closing slides provided for each.

To carry off gases of combustion from lamps of magnetographs, 3-inch copper ventilators are provided.

Entrance into building is had through a vestibule on the south side, as shown in Fig. 12, of 10 feet by 13 feet 8 inches outside dimension. Walls of entrance are built similarly to those of main building without the air shaft and but 2 feet of sawdust packing. The outside door can be closed before opening a second door leading into a small entrance hall, which is 6 feet wide and 11 feet long; from this room a third door leads into an opening in the sawdust packing, whence a fourth door opens into the passageway around the rooms. In placing these doors particular care was taken to make them close fitting. Entrance into either of the magnetograph rooms is to be had only from the hall between the two rooms through 8-inch refrigerator patterned doors packed with sawdust.

The diurnal change of the temperature has thus been reduced to a matter of a few tenths of a degree, and in fact it is believed that even this small variation will be eliminated as soon as some other source of light than the present lamps has been introduced. It has been repeatedly found that any sudden change of temperature which may amount to 50°–60° F. outside only makes itself felt gradually inside, and then does not amount to much over 0°.5, and may be even less than this amount. The annual range has

FIG. 13.—ESCHENHAGEN MAGNETOGRAPH AT COAST AND GEODETIC SURVEY MAGNETIC OBSERVATORY, BALDWIN, KANSAS.

PRINCIPAL FACTS OF THE EARTH'S MAGNETISM. 59

been converted into a gradual progressive change, for which allowance can easily be made, and amounting to between one-half and one-third of what it would be outside.

Fig. 13, which shows the magnetograph of the Eschenhagen pattern in place at the Baldwin Observatory, will exhibit the precise arrangement of the instruments.

The two instruments on the left are the declination variometer, by means of which the variations in magnetic declination are obtained, and the horizontal intensity variometer (one in the middle of view) giving the changes in horizontal component of the Earth's magnetic force. The magnets in both instruments are laminar pieces of well-hardened watch-spring steel, about an inch long (25^{mm}) and about one-fourth of an inch wide and about one-sixty-fourth of an inch thick—quite a difference from the meter-long magnets used in Gauss's time. The magnets are suspended by fine quartz fibers passing through the glass suspension tubes, and swing in copper damping boxes. The magnet in the declination instrument hangs in the magnetic meridian, whereas in the horizontal intensity instrument the magnet is turned at right angles to the magnetic meridian by means of torsion of the quartz fiber. A third instrument for registering the variation in vertical intensity completes the set.

On the right of the view are shown the lamp and the recording apparatus. A spot of light supplied by the lamp falls on the mirrors attached to the magnets, and is reflected onto the drum or cylinder inside the recording apparatus, a sensitized sheet of paper (platinum bromide) 15 by 52^{cm} being wrapped around the drum and the drum revolving either once in twenty-four hours or once in two hours, according to circumstances. As the magnet swings to and fro, the spot of light passes back and forth on the sensitized sheet, producing a curved or devious line full of peaks and hollows during time of magnetic disturbance. To provide a base line from which to count the changes, a second spot of light coming from a fixed mirror attached to each instrument traces its record on the revolving cylinder as a straight line.

A shutter operating automatically cuts off the light from the fixed mirror at intervals of one hour and thus the base line is interrupted, the distance between hourly breaks being about 20^{mm}, so that 1^{mm} of the base line represents 3 minutes of time, or 0.1^{mm} (the limit of reading), 18 seconds. If the drum revolves once in 2 hours, as it does during special work, then 1^{mm} of abscissa represents 15 seconds. One millimeter of ordinate, or $\frac{1}{25}$ of an inch, corresponds to a change of 1 minute in the magnetic declination, and about .000025 c. g. s. units in the horizontal intensity, or about $\frac{1}{8000}$ part of the absolute value of the horizontal intensity. As it is possible to estimate $\frac{1}{10}$ of a millimeter, the magnetograms will ordinarily be read to 0.1 of a minute and to .0000025 c. g. s. units ($\frac{1}{80000}$ H).

Figs. 14, 15, and 16 exhibit some of the interesting records already obtained. They are reproductions on half scale of the magnetograms obtained at the Cheltenham Magnetic Observatory with the Adie magnetograph. In this instrument each magnetic element (declination, horizontal and vertical intensity) is recorded on a separate photographic sheet, two days' record being obtained on each sheet. Each figure is composed of three sheets.

Fig. 14 is designed to show the character of the magnetic curves during a comparatively undisturbed period, and especially to exhibit the slight effect due to the Guatemalan earthquake. Beginning on top there are two curves—the declination curves—marked respectively April 18 to 19 and April 17 to 18, next two straight lines similarly dated, which serve as the base lines for the curves. From the explanation given in the preceding paragraphs it will be evident that the curves result from the spot of light coming from the mirror attached to the magnet, whereas the straight lines are due to the spot of light from the fixed mirror. Considering simply the curve and base line, each dated April 18-19, and measuring the perpendicular distances or ordinates between the base line and the curve at the hourly intervals marked, beginning with 5 p. m., April 18, passing through midnight and continuing until 4 p. m. of the following day, then the difference of these ordinates will give the changes in the magnetic declination from hour to hour for the period of time, 1^{mm} (one twenty-fifth of an inch) of ordinate on the original sheet being $1'.13$, and in the figures twice this amount, viz, $2'.26$. If the entire ordinate be converted into minutes of arc and added to the base-line value, the actual magnetic declination for each hour from April 18, 5 p. m., to April 19, 4 p. m., can be obtained. As the arrows on the side indicate, a rise in the curve means an increase of the declination (average value is about $5°.1$ west), whereas a fall in the curve means a decrease. *The hours as marked are for local mean time; to get seventy-fifth Meridian or Eastern standard time add 7.3 minutes.*

Thus at about 8 a. m., local mean time, the lowest value is reached, and between noon and 1 p. m. the highest one results, the total change amounting to 4^{mm}, or $9'.5$.

The same explanation will suffice for the next two curves (in the middle), the changes in the ordinates measured from the corresponding base lines giving the changes in the intensity of the horizontal component of the Earth's magnetic force or the force acting on the compass needle. The bottom curve and base line records the changes in the vertical intensity, the vertical intensity curve for April 17 to 18 having been omitted purposely to avoid confusion.

One millimeter of ordinate for either the horizontal or vertical intensity curve corresponds practically to 0.00005 c. g. s. unit, and on the original sheets to half of this amount. It will be noticed that the principal minimum of the horizontal intensity occurs at about 9 a. m. local mean time and the principal minimum of the vertical intensity curve occurs a little after 10 a. m.

Comparing the three separate sets of curves, it will be seen that the middle one—horizontal intensity—shows a number of small fluctuations not occurring in the other curves, and in fact this curve is rarely without disturbances of some kind.

Special attention is directed to the peculiar appearance of the curves (declination, horizontal intensity, and vertical intensity) between 9 and 10 p. m. on April 18, the curves being almost entirely obliterated for part of the way. This peculiar occurrence can be traced to the Guatemalan earthquake, the maximum effect of which was recorded at $9^h\ 40^m$ (seventy-fifth Meridian or Eastern time) on the Milne seismograph which Dr. H. F. Reid has had mounted at the Johns Hopkins University, Baltimore.

The late Professor Eschenhagen, who examined a number of such cases of earthquake effects registered on magnetic instruments, came to the conclusion that the effect was probably entirely a mechanical one, due to the vibration of the piers on which the instruments were mounted, and not a magnetic effect.

Other breaks in the curves, e. g., about 5 p. m., 8 a. m., and 4 p. m., are the "time breaks" and are purposely made in order to obtain the data for dividing up the base line into hourly intervals. (On the Eschenhagen magnetograph, as explained, this is done automatically.)

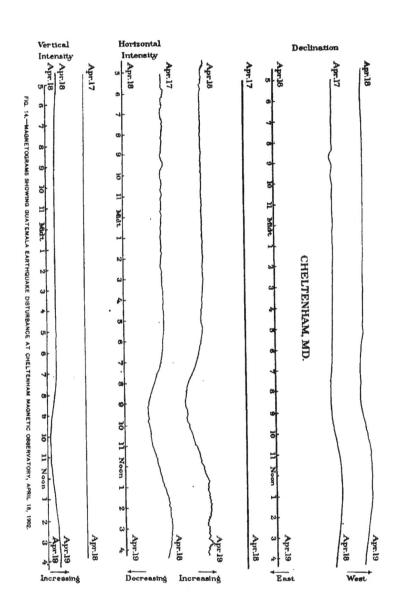

FIG. 14.—MAGNETOGRAMS SHOWING GUATEMALA EARTHQUAKE DISTURBANCE AT CHELTENHAM MAGNETIC OBSERVATORY, APRIL 18, 1902.

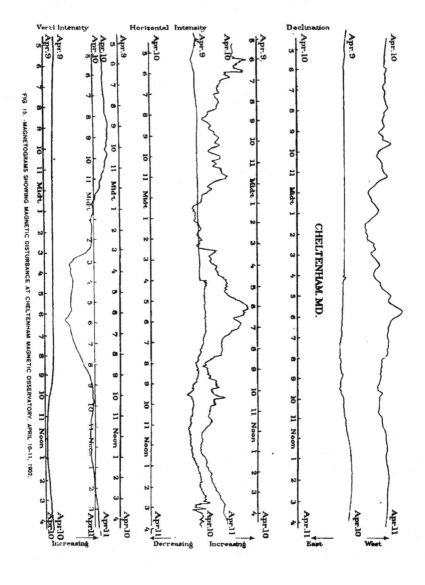

FIG. 15.—MAGNETOGRAMS SHOWING MAGNETIC DISTURBANCE AT CHELTENHAM MAGNETIC OBSERVATORY, APRIL 10-11, 1902.

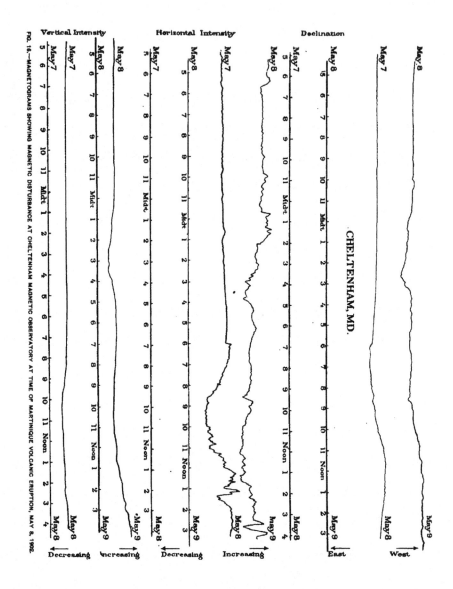

FIG. 16.—MAGNETOGRAMS SHOWING MAGNETIC DISTURBANCE AT CHELTENHAM MAGNETIC OBSERVATORY AT TIME OF MARTINIQUE VOLCANIC ERUPTION, MAY 8, 1902.

Fig. 15 shows the curves on a disturbed day. Looking at the second curve from the top, it is seen that the declination curve runs along smoothly until somewhat after 4 a. m., local mean time, April 10 (see second base line marked April 9-10), when it is suddenly interrupted. Thereafter it exhibits a number of fluctuations until the end of the curve. Continuing now on the upper curve marked April 10, still more marked fluctuations are exhibited until a little before 6 p. m., April 11 (first base line) the highest point is reached, the curve dropping thereafter. The change in declination between this point and the lowest one which occurred about four hours before is nearly 33'.

Passing on to the two middle curves—the horizontal intensity—it is found that the fluctuations are even more marked than for the declination curves, the beginning taking place very abruptly. The vertical intensity curve likewise exhibits large fluctuations.

This magnetic storm lasted about two days, and began practically simultaneously at the four magnetic observatories of the Coast and Geodetic Survey, viz, Cheltenham, Md.; Baldwin, Kans.; Sitka, Alaska, and Honolulu, Hawaiian Islands. At Sitka the disturbance in declination was 2° and over, part of the record being lost, having gone beyond the edge of the paper.

Fig. 16 reproduces a magnetic disturbance, which, as in the previous case, began very abruptly (see especially fourth curve). Now, the remarkable thing is this, that the time of beginning of this storm was coincident, as far as can at present be ascertained, with the time of the eruption of Mont Pelée (Martinique) on May 8. The magnetic disturbance began simultaneously at the Cheltenham and at the Baldwin observatories at $7^h 55^m$ St. Pierre local mean time. According to the newspaper reports, the catastrophe befell St. Pierre about 8 a. m. of May 8, and it was stated that the town clock was found stopped at $7^h 50^m$; how accurately this clock kept local mean time is, of course, not known. This disturbance was purely a magnetic one and not a seismic one, like that shown in Fig. 14, and was not recorded on seismographs. The Cheltenham magnetograms exhibit fluctuations amounting at times from .0005 to .0006 c. g. s. unit (about $\frac{1}{310}$ of the value of the horizontal intensity), and from 10' to 15' in declination.

On the morning of May 20, from $4^h 07^m$ to $4^h 16^m$ Eastern time, or $5^h 03^m$ to $5^h 12^m$ St. Pierre local mean time, there again occurred a slight disturbance of the magnetic needles at the Cheltenham Magnetic Observatory, beginning abruptly and reaching its maximum effect at $5^h 07^m$. From $11^h 57^m$ p. m., May 20, to $0^h 09^m$, May 21, Eastern time, or from $0^h 53^m$ to $1^h 05^m$ a. m., May 21, St. Pierre local mean time, a similar but somewhat larger disturbance occurred.

According to the cable dispatch from Governor L'Huerre, of the Island of Martinique (mentioned in the Associated Press dispatches), sent from Fort de France and dated Tuesday, May 20, the second eruption of Mont Pelée apparently began about $5^h 15^m$ a. m.—closely coincident with the time of the first magnetic disturbance given above.

Respecting the second magnetic disturbance, about midnight of the 20th, it is of interest to note that almost continuous earthquake shocks were felt at St. Augustine, Fla., from 9 to midnight, May 20.

The Coast and Geodetic Survey has undertaken a special study of the interesting occurrences above described, and has sent a request for information to every magnetic observatory in foreign countries.

MAGNETIC CHARTS.

Isogonic Lines.

The most convenient form in which to represent magnetic data for practical use, namely, by drawing lines through the places having the same magnetic declination the same magnetic dip, or the same magnetic intensity, is generally supposed to have been employed for the first time by Dr. Edmund Halley, the noted astronomer, who, at the beginning of the eighteenth century, published two charts of the "lines of equal magnetic variation (declination)," which are here called, respectively, the "Atlantic Chart" and the "World Chart." According to Hellmann, however, Christoforo Borri, of Milan, appears to have made the first attempt to construct lines of equal magnetic declination, but did not publish them.

The "Atlantic Chart," doubtless published in 1701, gave the lines of "equal magnetic variation" chiefly over the Atlantic Ocean, based upon Halley's observations, made between 1698 and 1700 on the ship *Paramour Pink*, the expenses of the expedition having been borne by the English Government, this fruitful expedition representing the first systematic effort made at a magnetic survey of the globe. In no case were the lines on this chart drawn over land areas.[a]

The "World Chart," frequently referred to under the title of "Tabula Nautica," published later than the preceding one (probably in 1702), besides containing the "lines of equal variation" for the Atlantic Ocean, also gave them for the Indian Ocean and the extreme western part of the Pacific Ocean. (See Fig. 18.) In a few instances the lines were drawn across the continents. This was reproduced by Airy in the Greenwich Observations for 1869, and again by Hellmann[a] in 1895.

Revisions of Halley's chart, made necessary by the progressive change in these lines of equal magnetic declination with the lapse of time, were made after Halley's death by Messrs. Mountaine and Dodson in 1744 and 1756. The most complete collection of early charts of the lines of equal magnetic declination (isogonic charts) and of equal magnetic dip (isoclinic lines) will be found in Hansteen's Atlas, belonging to his famous work "Magnetismus der Erde," Christiania, 1819, and in Hellmann's facsimile reprints,[b] to which latter work the reader is referred for a detailed historical account.

The following series of isogonic charts from 1600 to 1858 (Figs. 17–20) have been reproduced on a reduced scale from Neumayer's excellent Atlas des Erdmagnetismus, published by Justus Perthes, Gotha, 1891, those of 1600 and 1800 being due to Hansteen, and published in 1819, that of 1700 to Halley, and the one of 1858 to the British Admiralty. Van Bemmelen has recently constructed isogonic charts for 1500 (see Fig. 4), 1550, 1600, 1650, and 1700, based on an exhaustive collection of early declination values.[c] A careful scrutiny of them is earnestly recommended to the reader. Let him pick out, for

---

[a] A copy of this chart, whose existence had escaped attention, was found by the writer in 1895 in the British Museum, and reproduced by him with commentary notes in the journal "Terrestrial Magnetism," Vol. I, No. 1, 1896.

[b] "Die ältesten Karten der Isogonen, Isoklinen, Isodynamen," Berlin, A. Asher & Co., 1895. (At the time of this publication Hellmann was not aware of the "Atlantic Chart," and so erroneously believed that the "World Chart" was the original Halley Chart of 1701.)

[c] "Die Abweichung der Magnet Nadel," Batavia, 1899.

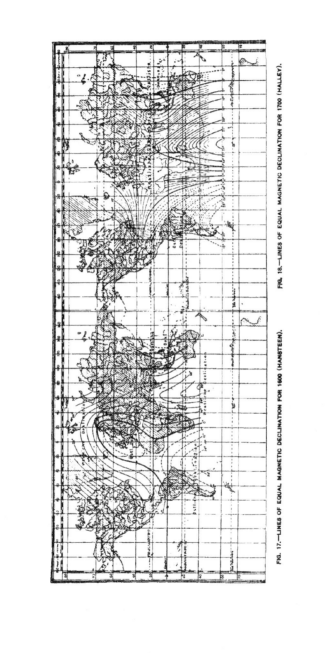

FIG. 17.—LINES OF EQUAL MAGNETIC DECLINATION FOR 1600 (HANSTEEN).   FIG. 18.—LINES OF EQUAL MAGNETIC DECLINATION FOR 1700 (HALLEY).

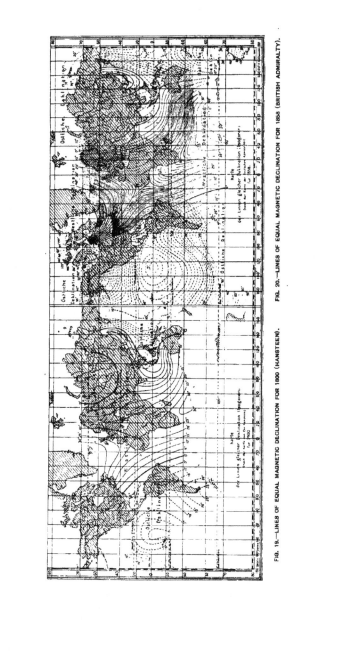

FIG. 19.—LINES OF EQUAL MAGNETIC DECLINATION FOR 1800 (HANSTEEN).   FIG. 20.—LINES OF EQUAL MAGNETIC DECLINATION FOR 1858 (BRITISH ADMIRALTY).

example, an agonic line (line of no magnetic declination) and follow its various transformations from period to period. Or let him compare the chart of 1600 with that of 1905, given later, and notice what a complete reversal there has been in the distribution of the Earth's magnetism, as represented by the lines of equal magnetic declination. Thus in 1600 the declination over the western and southern parts of the Atlantic Ocean and over western Europe and western Africa was *east*, whereas to-day, over the same portions of the Earth, the declination is *west*.

The chart (Fig. 21) of 1905 was reproduced from the British Admiralty Manual of Deviations of the Compass, 1901. The isoclinic chart, giving the lines of equal magnetic dip for 1905 (Fig. 22), has been taken from the same source.

In looking over the series of isogonic charts, two main lines of zero or no magnetic declination (agonic lines) intersecting the equator, a western one and an eastern one, can be recognized. If the longitudes of the intersections were determined from time to time and represented graphically, the ordinate being the longitude and the abscissa the corresponding year, it would be seen that for nearly three hundred years there has been a progressive and almost uniform motion of these two agonic lines to the *westward*, the western agonic at an average annual rate of nearly 14 minutes in longitude and the eastern agonic at an average annual rate of about 8 minutes in longitude. Were the western agonic to make a complete revolution of the Earth at the rate given, it would take it nearly one thousand six hundred years, whereas the eastern agonic would require about two thousand six hundred years. These results show how fruitless it is to endeavor to determine the secular change period from the supposed motion of the agonic lines around the Earth. The result reached will depend not only upon the agonic selected, but also upon the parallel of latitude along which the sliding around the Earth is supposed to take place.[a]

MAGNETIC MERIDIANS.

The lines of equal magnetic declination, while representing magnetic declination data in a convenient and practical form, do not actually exist in nature; they are merely an artificial set of lines devised to serve a useful purpose, which they admirably fulfill. The so-called "magnetic meridians," with which the isogonic lines are often confounded, give a better representation of the actual magnetic condition of the Earth. They are the lines along which one would travel were he to set out at any place on the Earth and always follow the direction of the compass needle, and hence they exhibit at every point the actual direction of the compass needle, not by numbers, but by angles. The magnetic declination at any point will be the angle between the magnetic and the true meridian passing through the point.

Fig. 23 gives the magnetic meridians for 1836 as drawn by Captain Duperrey. It will be noticed that they all pass through two points—one in the Northern Hemisphere, the North Magnetic Pole, and the other in the Southern Hemisphere, the South Magnetic Pole. The lines cutting across the magnetic meridians at right angles, which in the present instance are the lines of equal "potential," Duperrey termed the "magnetic

---

[a] In this way Lord Kelvin deduced his much-quoted period of nine hundred and sixty years.

64   PRINCIPAL FACTS OF THE EARTH'S MAGNETISM.

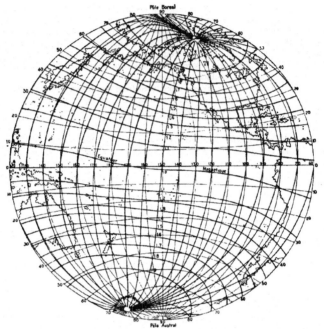

FIG. 23.—Magnetic meridians for 1836 (Duperrey).

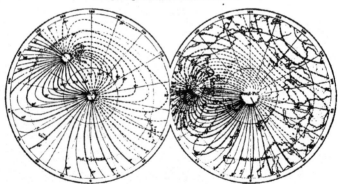

FIG. 24.—Lines of equal magnetic declination in the polar regions for 1885 (Neumayer).

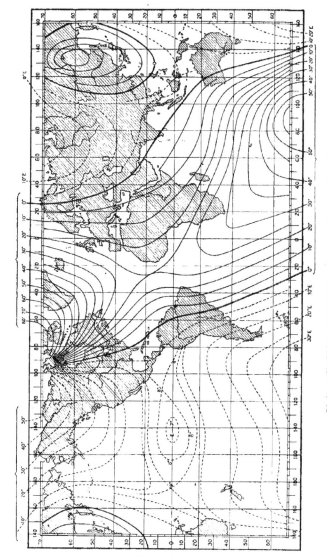

FIG. 21.—LINES OF EQUAL MAGNETIC DECLINATION FOR 1905 (BRITISH ADMIRALTY).
[Full lines indicate west magnetic declination and broken lines east magnetic declination.]

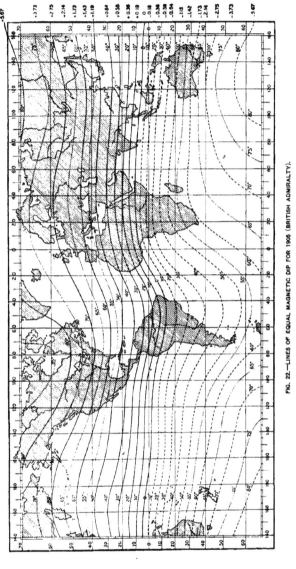

FIG. 22.—LINES OF EQUAL MAGNETIC DIP FOR 1905 (BRITISH ADMIRALTY).

parallels." It is more usual, however, to call the lines of equal dip the "magnetic parallels."

The isogonic lines, on the other hand, pass through four points—two in each hemisphere, the magnetic pole and the geographical pole. (See Fig. 24, which gives these lines for 1885, as reproduced from Neumayer's "Atlas.") In other words, at both points in each hemisphere it is possible to have all values of the magnetic declination; at the magnetic pole because there all magnetic meridians converge, and at the geographical pole because there all true meridians meet, and since the magnetic declination is the angle between the magnetic meridian and the true meridian, it is therefore possible to have every value of the magnetic declination at each of the two points. There is this distinction between them, however: At the magnetic pole the compass has no definite direction, all the force acting at this point being vertically downward, so that there is no force in the horizontal plane pulling the compass into any fixed direction; the true north and south direction is, however, a fixed one. At the geographical pole, however, the true direction is anything you please, while the compass direction is a perfectly definite one.

MAGNETIC SURVEYS.

GENERAL REMARKS.

The present time is witnessing a remarkable revival of interest in magnetic work. Magnetic surveys are either already under way or contemplated for the near future in nearly every civilized country. From the Antarctic expeditions valuable results may be expected in a region almost destitute of data, and where in fact nothing has been obtained since the observations of Ross and Crozier in the ships *Erebus* and *Terror* and of Moore and Clerk in the *Pagoda*, sixty years ago.

Unfortunately, however, in the regions of the Earth where information regarding the magnetic needle is of the highest practical importance to the seaman in these days, when every effort is bent to increase the speed of a vessel by a knot over the great ocean basins continually traversed—the Atlantic, the Pacific, etc.—there almost no magnetic data are at present being obtained. But very little data regarding the magnetic declination of the needle, to say nothing of the dip and intensity, have been obtained in the ocean areas since the advent of iron ships, except from occasional expeditions. The present lines of equal magnetic declination, or, as the mariner terms them, "lines of equal magnetic variation," over these waters depend almost entirely upon results acquired in wooden ships 50 to 100 years and more ago. It is therefore impossible to state just how accurate they may be. When it is remembered that in times of fog and darkness, with no celestial object visible, sole reliance must be placed on the log, compass, and the "variation" charts, the importance of a systematic magnetic survey of ocean areas needs no further argument. Fortunately all evidence goes to show that over the deep waters of the ocean most frequently traversed, the Atlantic, the present lines of equal magnetic declination are doubtless correct within 1°. In shallow waters, however, and near coast lines, where danger of shipwreck is most imminent, greater errors in the lines can be confidently expected. Respecting the Pacific Ocean, it is impossible to form an accurate opinion as to the correctness of the mariner's "variation charts." Unfortunately, almost the universal practice employed by seamen in these waters is to deduce the compass deviation, or compass correction, due to the ship's own magnetism, entirely with the aid of the "variation charts;" and rarely do they control their table of deviation corrections by "swinging ship." The difference between the observed "variation of the compass" on board ship and that scaled from the variation charts, is ascribed wholly to the local magnetism of the ship, and called the "deviation of the compass" on the course on which the "variation" was observed. This difference, however, is due to three causes: (1) Ship's own magnetism; (2) error in variation charts; (3) error in mariner's observation. If mariners in the Pacific Ocean would likewise swing ship, when opportunity offered, and thus determine the deviations of the compass on various courses independently of the charts, valuable data would be furnished those whose duty it is to construct "variation charts."

The Coast and Geodetic Survey is making arrangements to fit out its vessels with the necessary instruments for determining the magnetic elements at sea.

Besides the need of a systematic magnetic survey of ocean areas, there are vast areas of the Earth, some under the control of civilized nations, which have not yet been magnetically explored.

The complete solution of some of the vexed problems of the Earth's magnetism of international interest, can not be accomplished until some of the gaps in knowledge as above pointed out have been filled.

The necessity of obtaining facts for keeping "variation charts" up to date, i. e., correcting them for secular change, has already been made apparent in the previous section on "Magnetic charts." It has been shown that it does not take many years to make appreciable changes. Fig. 25, due to Neumayer, gives the amount of annual change of the magnetic declination for various parts of the Earth. It will be seen that along the tracks usually followed by steamers plying between New York and England, the change may be as much as 6 minutes a year—that is, 1° in 10 years—while over other ocean areas, e. g., South Atlantic, a change of 1° may be expected in about 6½ years.[a] Over the greater part of the Pacific Ocean, the change, at present (it may not always be so), is on the average about 2 minutes per year, or 1° in 30 years. For the United States, as will be seen by turning to the Chart, the change is from 0' to 5', the average being about 3', or 1° in 20 years.

---

[a] An error of 1° in a course a mile long amounts to an error in distance of 92.2 feet. Supposing a speed of 20 knots an hour, a vessel persisting on a course erroneous by 1° would be out at the end of the day's run about 9.6 miles, or 8.4 knots—nearly one half hour in time. Thus, when every nerve is being strained to cut down the time of crossing the ocean by an hour or so, the need of being able to hold a vessel true to her course is apparent.

When the mariner is obliged to rely entirely upon the compass and the log, the uncertainty in fixing the ship's position at the end of a day's run is due to the error in distance traversed and to the error resulting from imperfect knowledge of the true bearing of the course followed. If, therefore, it were possible to add another factor for fixing the ship's position, e. g., if sufficiently accurate dip observations were possible on board ship, they might help materially, with the aid of the isoclinic charts, to fix the position.

In times of clear weather, when celestial objects are visible, there would of course be no need of a "magnetic" method for determining the ship's position, but when no astronomical method can be employed then any additional information to that supplied by the compass and the log is greatly to be desired.

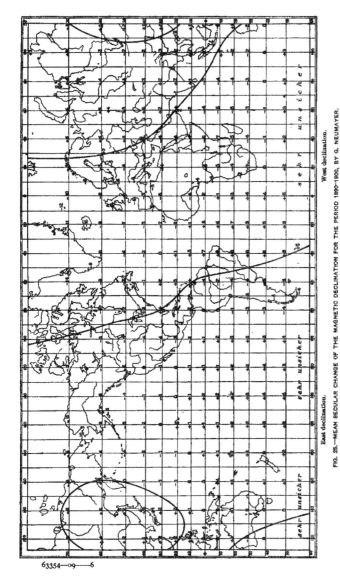

FIG. 25.—MEAN SECULAR CHANGE OF THE MAGNETIC DECLINATION FOR THE PERIOD 1890-1900, BY G. NEUMAYER.

[The plus sign denotes that the prevailing declination, whether it be west or east, is increasing by the amount given in minutes, whereas the minus sign means that it is decreasing.]

HISTORICAL SUMMARY.[a]

The first complete magnetic survey in which the three magnetic elements—declination, dip, and intensity—were determined, and which was executed as a national piece of work and was coextensive with the limits of the country surveyed, was that of the British Islands, corresponding to January 1, 1837. This survey was undertaken in 1836 at the request of the British Association for the Advancement of Science, and was completed in 1838. The example set by Great Britain was speedily followed by the execution of similar surveys in various portions of the globe—in Austria, Bavaria, Germany, Holland, Belgium, France, Canada, etc. At the present time nearly every civilized country has been surveyed magnetically to a greater or less extent.

But such surveys must be repeated after the lapse of a number of years on account of the slow, yet appreciable, change forever going on in the Earth's magnetic state, which change, as one of the most noted physicists has truly said, is a warning "that we must not suppose that the inner history of our planet is ended." Thus after the lapse of twenty years Great Britain—again at the instance of the British Association for the Advancement of Science[b]—repeated its original magnetic survey.[c] The observations were taken between 1857–1862. In the Philosophical Transactions of the Royal Society for 1870 will be found a full account of this survey and likewise of the earlier one. In this paper Sir Edward Sabine combined the observations of the two surveys and drew the isomagnetic lines for the mean period of 1842–1845. Recently Great Britain has completed a *third* magnetic survey, far more elaborate than any of the preceding surveys. This survey, one of the most carefully executed up to date, was conducted by two most eminent physicists, Professors Rücker and Thorpe.[d] It is a most fruitful piece of work. Observations of the three elements were made at first by the two distinguished professors themselves between the years 1884–1888 at 205 places.[e] The government grant committee of the Royal Society then made a liberal grant so that the survey might be carried out on a larger scale than hitherto attempted. Two assistant observers were then employed, and with their aid, in the four years 1889–1892, the grand total of the number of stations was brought up to 882, making on the average 1 station to every 139 square miles of land area.[f] The isomagnetic lines corresponding to the epoch 1886–1890, and based on the 205 observations made between the years 1884–1888, were drawn, and likewise those as based on the 677 stations observed in 1889–1892, were constructed for the epoch 1891, and finally the

---

[a] Quoted largely from the writer's First Report on Magnetic Work in Maryland.

[b] Doubtless no national organization has done so much for the advancement of the subject of terrestrial magnetism as this most distinguished body of scientific men. Money grants have been freely made; committees on terrestrial magnetism composed of the most eminent physicists have been formed from time to time, and cooperation has been extended and encouragement given to magnetic enterprises in many ways.

[c] Report on the Repetition of the Magnetic Survey of England, by Maj. Gen. Edward Sabine. Report of the British Association for the Advancement of Science for 1861.

[d] Dr. Thorpe has made a number of determinations of the magnetic elements in the United States.

[e] The results were published in the Phil. Trans. of the R. S., 1890, A, p. 53, the memoir constituting the Bakerian lecture of that year.

[f] The results of this last work have just been published, Phil. Trans. R. S., vol. 188, A, 1896.

lines as resulting from *all* the stations were obtained. A splendid opportunity was thus afforded for testing the accuracy with which the positions of the isomagnetic lines, e. g., the lines of equal magnetic declination or variation, can be inferred from observations in greater or less number. For further details the reader is referred to Professor Rücker's interesting account[a] published in Terrestrial Magnetism for July, 1896.

Professor Rücker's[b] results regarding the relation of magnetic disturbances and geological formations are of such universal interest that they are quoted in toto:

"It has long been known that just as the secular variation of the magnet is accompanied by minor diurnal changes, so the large alterations in the direction of the compass and dipping needle, which are observed when we move from place to place on the surface of the earth, are affected by irregularities which are due to purely local causes. Thus the declination is greater in Ireland than in England; but the increase is not uniform as we pass from one country to the other. In fact in some districts an abnormally large increase is followed by a decrease.

"These curious inequalities must be due to local disturbing forces, and the large number of observations which have been made in this country [Great Britain] have enabled us to determine with more than usual accuracy the magnitude and direction which the magnetic forces would assume if they were undisturbed by any local cause, and from the difference between things as they then would be and things as they actually are, we can calculate the magnitude and direction of the disturbing forces themselves. When these are represented on a map, it is found that there are large districts of the country in which the disturbing horizontal forces act in the same direction; in one region the north pole of the needle will be deflected to the east, in another to the west, and, as we pass from one of these districts to the other, we always find that at the boundary the downward vertical force on the north pole of the needle reaches a maximum value. We are thus able to draw upon the map lines toward which the north pole of the needle is attracted. It is found that the exact position of these can be determined with considerable accuracy, and that the lines can be traced without any possible doubt through distances amounting, in some instances, to a couple of hundred miles. The key to this curious fact is probably furnished by observations in the neighborhood of great masses of basalt or other magnetic rocks. If these were magnetized by the induction of the earth's magnetic field, the upper portions of them would, in this hemisphere, attract the north pole of the needle; and it is found that where large masses of basalt exist, as in Antrim, in the Scotch coal fields, in North Wales, and elsewhere, the north pole of the needle is, as a matter of fact, attracted toward them from distances which may amount to 50 miles. The thickness of the sheets of basalt is in most cases too small to furnish a complete explanation of the observed facts, but it is quite possible that these surface layers of magnetic matter are merely indications of underground protuberances of similar rocks from which the surface sheets have been extruded. At all events, there is no possible doubt of the fact that where large masses of basalt occur, the north pole of the needle tends to move toward them.

"There are other regions where the attractions are manifest, but where, nevertheless, no magnetic rocks occur upon the surface; but it is most probable that the cause is the same, and that it is due to the mere accident of denudation that in one case we can, and in the other we can not, point to the magnetic rocks to which the anomalous behavior of the compass is due. *If this be so, it is certainly interesting that magnetic observations should enable us to penetrate to depths which the geologist can not otherwise reach, and that the lines which we draw upon the surface of the map, as those to which the north pole is attracted, may, in fact, roughly represent the ridge lines of concealed masses of magnetic rocks, which are the foundations upon which the deposits studied by the geologist have been laid.*

---

[a] A. W. Rücker: A Summary of the Results of the Recent Magnetic Survey of Great Britain and Ireland conducted by Professors Rücker and Thorpe:
   I. On the Accuracy of the Delineation of the Terrestrial Isomagnetic Lines.
   II. On the Accuracy of the Determination of the Local Disturbing Magnetic Forces.
   III. On the Relation between the Magnetic and the Geological Constitution of Great Britain and Ireland.
[b] Extracted from Terrestrial Magnetism, Vol. III, pp. 42-43.

"There is some ground for thinking that if these great underground wrinkles exist, they have affected the rocks which are superposed upon them, especially those which are of a comparatively early date. As a general rule, if older rocks appear in the midst of newer ones, the pole of the magnet will be attracted toward the protruding mass; but this rule holds good only of the rocks of Carboniferous or Pre-Carboniferous age, and does not apply to later deposits. As a striking example, I may state that the Pennine Range—which is sometimes called the "backbone of England"—is a mass of millstone grit rising amid younger rocks Down this a well-marked magnetic ridge line runs. Similarly, in the neighborhood of Birmingham, the Dudley and Nuneaton coal fields are surrounded by more modern deposits. A curious horseshoe-shaped ridge line connects these two, and then runs south to Reading, which is, magnetically speaking, one of the most important towns in the Kingdom. East and west from Dover to Milford Haven, and then across the English Channel to Wexford, runs a ridge of the older rocks, called by geologists the Paleozoic ridge, concealed in many places by newer deposits. Hollowed out in this are the South Wales and Forest of Dean coal fields, and in another hollow within it lies the coal which has recently been discovered at Dover. Closely following this protruding mass of the older rocks is a magnetic ridge line which passes through Reading, and thus we have a magnetic connection between the anticlinals of Warwickshire and the Paleozoic ridge. From the neighborhood of Reading also another magnetic ridge line runs southward, entering the channel near Chichester. M. Moureaux, who, with most untiring energy, has for many years been investigating, single handed, the magnetic constitution of France, has discovered the continuation of this line on the French coast near Dieppe, and has traced it through the north of France to some 50 miles south of Paris. The energy which is now being displayed by magnetic surveyors in many countries will, no doubt before long, prove that the network of these magnetic ridge lines is universal, and the relations between them and the geological conformation of the countries in which they lie will be so studied that our inductions will be based upon an adequate knowledge of facts."

To give an intelligent and fair account of all work done in recent years in this special field of human activity would require far more space and time than is available. On the European continent, in nearly every country, elaborate magnetic surveys are either at present in progress or have just been finished or are in contemplation. One of the most detailed in recent years is that of Holland, by Dr. Van Rijckevorsel, for the epoch 1891.0, embracing 278 stations over an area about equal to that of Maryland, or averaging about one station to every 40 square miles. This survey of Holland is especially interesting from the fact that though it was made over an area *superficially* destitute of striking geological features, it nevertheless revealed marked disturbances. The author sums up his conclusions thus: "Little even as we know about the geology of the Netherlands, the magnetic maps must bring every one to the conviction that in some cases, in many perhaps, there must be a direct relation between geology and terrestrial magnetism, and that many of the magnetic features must be in some way determined by the geological structure of the underground. What these geological features might be we are at present unable to tell. What kinds of rock may be hidden at a depth of 300 meters or more under the peat bogs and heaths of the Netherlands, and the clay, sand, and pebbles immediately underlying these, we do not know—rocks which, although under ground, are yet perhaps in some places so near the surface as to be an effective barrier against the inroads of the sea, which has fair play in other districts."

So, likewise, important and interesting results were obtained by Professor Liznar, who conducted the magnetic survey of Austria. The magnetic survey of most of the German States (for a second time, and on a more elaborate scale than during Lamont's time) is now in progress. The Russian Government has been planning a magnetic survey of its extensive domains, and it is hoped that the funds will soon be forthcoming.

Magnetic surveys of India and of Egypt are being undertaken under the auspices of the English Government. Instances could be multiplied in which good and important work has been accomplished by magnetic surveys, as, for example, in France, Italy, Belgium, Denmark, Norway and Sweden, etc.

Extremely interesting investigations in the greatly disturbed areas in Russia between Kursk and Odessa have been made by Leyst, Moureaux, and Passalskij.[a]

Enough has been given, however, to show that by undertaking similar work the United States is simply keeping in touch with a general movement that is manifesting itself most actively in the civilized world to-day. It is recognized that in the eager and impatient endeavor to unravel the mysteries of the celestial regions the terrestrial mysteries, manifesting themselves every moment, have been woefully neglected. The science of our Earth is still in its infancy, and the astronomer has been made painfully aware of the fact that more attention must be given to the study of the physical history of the planet. There is every evidence that a reaction in scientific thought in this regard has set in that is bound to grow, and it is most desirable that the United States should keep in line with this onward movement.

MAGNETIC SURVEY OF THE UNITED STATES.

In concluding, brief reference to the history of terrestrial magnetism in the United States is made, so that one may form some opinion as to the place to be ascribed to this country in the development of magnetic surveys.

The earliest attempt at a detailed State magnetic survey appears to have been made by Prof. Alexander Dallas Bache in 1840-1843, just before he was called to the Superintendency of the Coast Survey. He called his survey a "Magnetic survey of Pennsylvania and parts of adjacent States." Observations were made at 22 points within Pennsylvania; they did not in every case embrace the three elements. Professor Bache made these observations during his summer vacations from 1840-1843 and at private expense.

When Bache became Superintendent of the Coast Survey magnetic work was incorporated in the work of the Survey. Since then magnetic observations have been made in every State of the Union by the Coast Survey, and the drawing of isomagnetic maps and the furnishing of the data for allowance for the secular change have become regular authorized functions of the Survey. The extension of the observations in such manner as would fulfill the modern requirements of a magnetic survey could not be undertaken until 1899, when the United States Congress, acting upon the recommendation of Dr. Henry S. Pritchett, then Superintendent of the Coast and Geodetic Survey, largely increased the appropriation for magnetic observations over what it had been before that date.

An officer of the Survey was placed in immediate charge of the details of the work in the field as Inspector of Magnetic Work, a division of Terrestrial Magnetism was created in the Office of the Survey, and operations were extended to the limit fixed by the amount of money available.

Magnetic observations, more or less complete, and magnetic tours, more or less extensive, had been made previous to Bache's work, referred to above, e. g., by Long

---

[a] See Terrestrial Magnetism, Vol. IV, p. 235, and Vol. VII, No. 2.

(1819), Nicollet (1832-36), Locke (1838-43) and Loomis (1838-41). The last made the first general collection of magnetic observations for this country and has the honor of having drawn the first magnetic maps. To be sure, these maps, covering the eastern part of the United States, owing to the scantiness of the material, were only rough approximations; nevertheless, when, sixteen years later, a more complete map was made by the Coast Survey, Professor Bache declared that between his own map and that of Loomis, when proper allowance was made for the secular change, the "agreement was remarkable." This epoch of about 1840 is remarkable for the number of zealous and devoted students of terrestrial magnetism among the famous scientists in the United States. It is hoped that before long some of the physicists of this country can again be counted in the list of eminent magneticians.

From 1878-1883 Prof. Francis E. Nipher, professor of physics at Washington University, St. Louis, undertook a detailed magnetic survey of Missouri. Professor Nipher must be duly credited with the spirit and enterprise he exhibited in the inauguration of this survey. He was dependent entirely upon private aid for the defraying of the expenses of the work. The instruments were loaned by the Coast and Geodetic Survey. Professor Nipher has published thus far five reports of this work[a]. He has, however, not been able to complete the survey, and so no final report and no maps have been published. He had observed, with the aid of assistants, at 149 stations, or on the average at one station to 438 square miles.

At the same time some preliminary observations appear to have been made by Prof. Gustav Hinrichs, in Iowa, but the survey does not seem to have progressed far beyond a beginning.

Next comes the declination survey carried out—this time under State auspices— under the direction of Prof. George H. Cook, then State geologist of New Jersey, now deceased. The period of the survey was 1887-1890, all but a few of the observations to the extent of 158 stations having been made within this period. There was thus on the average one declination station to about 52 square miles. The observations were not made with special magnetic instruments, but good surveying transits were used. The observers appear to have executed the work as carefully as the methods and instruments would permit.

It was a commendable piece of work, as far as it went, but it was not complete. In order to derive the full benefit from magnetic surveys, it is absolutely essential to determine not declination alone, but also dip and intensity. Experience has repeatedly shown that with proper instruments a skilled observer can determine the three magnetic elements at practically the same cost in money and time as when declination results alone are determined. The principal expense and labor occurs in getting to a station and determining the *true* meridian. After this, the magnetic work, with proper instruments and modern methods, can be expeditiously and economically performed.

In 1896 the State Geologist of Maryland, Prof. W. B. Clark, inaugurated a magnetic survey of Maryland, and intrusted it to the writer. The work was done principally in the summer months of 1896 and 1897 and in the spring of 1899, the expense being borne by the State of Maryland, except in 1899, when the expenses were divided between the State and the Coast and Geodetic Survey. In 1900 all of the expenses

---
[a] Transactions of the St. Louis Academy of Sciences, 1878-1886.

were borne by the Coast and Geodetic Survey, and the instruments used throughout the work (1896-1900) were loaned by this Survey. This work has resulted in giving Maryland the most detailed magnetic survey of any State, there being on the average one station to about 100 square miles. Holland is the only country which excels Maryland in this respect, having, on the average, one station to every 45 square miles. The results have been published in two Reports by the Maryland Geological Survey. A number of interesting facts have been revealed, especially in the locally disturbed areas; it has been amply demonstrated that if the geologist desires to invoke the aid of magnetism in the solution of some of the vexed problems with reference to subterranean formations at depths impenetrable by ordinary means, he must use approved magnetic methods, and not be content with instruments which admit of simply "ordinary" accuracy.

A magnetic survey of North Carolina was made between 1898 and 1899, by J. B. Baylor, of the Coast and Geodetic Survey, under the joint auspices of the Coast and Geodetic Survey and the North Carolina Geological Survey (Prof. J. A. Holmes, State Geologist). The "General Report" of this work, prepared by Messrs. Baylor and Hazard, will be found in Appendix 9, Coast and Geodetic Survey Report for 1898-99. (See also Bulletin No. 41.)

As stated above, since 1899 the Coast and Geodetic Survey has been enabled to undertake systematically a magnetic survey of the territory under the jurisdiction of the United States.[a] The general plan on which the magnetic survey is being conducted has been published in Appendix 10, Coast and Geodetic Survey Report for 1898-99.

It is the intention to make first a general survey with stations about 25-30 miles distant and to occupy between 400-500 stations a year. After the general survey has been completed additional stations will be placed where most needed, as, for example, in the locally disturbed areas revealed by the general survey. Also, besides the continuous observations at the magnetic observatory stations, the magnetic elements will be redetermined at a number of well-chosen and uniformly distributed places from time to time, in order to determine the amount of secular change, and thus make it possible to keep the magnetic charts up to date. For fuller information refer to the Appendix cited.

---

[a] The areas of the countries at present belonging to or under the jurisdiction of the United States are, approximately, as follows:

	Square miles.
United States	3 026 800
Alaska	590 900
Philippine Islands	120 400
Hawaiian Islands	6 400
Porto Rico	3 400
Guam Island, Tutuila Island, and Midway Islands	300
Panama Canal Zone	470
Total	3 748 670

The area controlled by the United States is equal to that of Europe, or about one-fifteenth of the entire land area of the globe.

THE EARTH'S MAGNETIC POLES AND MAGNETIC MOMENT.

Magnetic Poles.

The definition for the magnetic poles of the Earth commonly accepted, is that they are the points on the Earth's surface where the dipping needle stands precisely vertical, i. e., where the dip is 90°, at the north magnetic pole, the north end of the dipping needle pointing vertically downwards, and at the south magnetic pole, the south end of the same needle pointing vertically downwards. Excluding "local magnetic poles,"[a] caused by extraordinary local deposits of attracting masses, there are but two such points, one in the Northern Hemisphere and the other in the Southern Hemisphere; their approximate positions will be presently given, and it will be seen that they are not diametrically opposite each other. At these points, as all of the Earth's magnetic force acts vertically downwards, there is no horizontal component to act on the compass needle and hold it in any fixed direction, hence the compass needle at the magnetic poles, except for extraneous disturbing influences, remains in any position in which it may be placed.

The points of greatest intensity of the total magnetic force, because of the irregular way in which the Earth is magnetized, are not coincident with the magnetic poles as above defined; barring out local manifestations there will be found to be four such points, two in each hemisphere, termed the "foci of greatest magnetic intensity." The stronger of the two intensity foci in the Northern Hemisphere, was, according to Lefroy's observations in 1843–44, in latitude 52° 10′ north and in longitude 99° 59′ west of Greenwich, hence somewhat south of Hudsons Bay and considerably south of the north magnetic pole.

It can not too clearly be pointed out that the points on the Earth's surface termed "magnetic poles," are by no means to be compared to the poles of a bar magnet. If they were similar in their action, then, manifestly, the weight of iron particles ought to increase enormously with approach to the magnetic poles. This, however, is known not to be the case. The increase in the weight of iron as the pole is approached, on account of the increase of the vertical force of the Earth's magnetism, would only be about one-eighth of that due to the well-known increase of gravitational force ($\frac{1}{280}$) because of the flattening of the Earth at its rotation poles. The Earth is a *spherical* magnet, and not at all to be likened to a bar magnet. A bar magnet at the center of the Earth which would produce the magnetic facts observed on our globe would have its magnetic poles practically coincident with each other. Two well-known investigators, Kraft and Biot, found that the nearer to each other they assumed the poles of a fictitious bar magnet placed at the center of the Earth to be, the closer the correspondence between their computed results on this hypothesis and the observed facts; so that the "equivalent magnetic poles of a spherical magnet" are practically the same distance from all points on the Earth's surface, and this accounts for the very slight increase in the weight of iron which might be expected if it were carried from the "magnetic equator" to the "magnetic pole."

---

[a] A "local magnetic pole" was found by Messrs. Leyst and Moureaux in Russia, between Kursk and Odessa; the writer in the fall of 1900 found one near Juneau, Alaska, viz, on Douglass Island, opposite Sheep Creek. There are a number of such "local" poles.

Hence there are no points on or near the Earth's surface equivalent in their action to the poles of a bar magnet; the points which are termed the "magnetic poles of the Earth" are simply the points of intersection of the direction of vertical dip with the Earth's surface.

If the Earth were uniformly magnetized throughout instead of being heterogeneously magnetized, the line joining the "equivalent magnetic poles," if prolonged, would pass through the points on the Earth's surface where the dip is equal to 90°, and this line would be the "magnetic axis" of the Earth. Only about seven-tenths of the total force of the Earth's magnetism can be referred to a homogeneous magnetization, the remainder being due to irregular magnetizations. Hence we must expect neither that the points of vertical dip lie diametrically opposite to each other nor that the magnetic axis of the uniform magnetization should coincide with the straight line connecting them. The said magnetic axis passes through the Earth's center and connects the points on the surface, lying respectively in latitude 78°.3 north, longitude 67°.3 west, and in latitude 78°.3 south, longitude 112°.7 east, while the straight line connecting the magnetic poles does not pass through the center of the Earth but off to one side.

In consequence of the heterogeneous magnetization of the Earth a magnetic meridian line is not a straight line leading to the magnetic poles, but a very devious line indeed. And thus a great circle passed through the direction pointed out by a compass needle at any given place will not pass through the magnetic poles, and the opposite intersections of two of such circles will not coincide with the magnetic poles.

It is desirable to advert to one more matter before proceeding to give the position of the "magnetic poles." Gauss defined these points as the places of minimum and of maximum potential, the former being the north magnetic pole. The points so defined would coincide with those of vertical dip, if no part of the Earth's magnetism be due to electric currents which pass from the air into the earth and *vice versa*. It would seem as though we have some indication that a small part (about 2 or 3 per cent) of the Earth's magnetic force is to be ascribed to such currents.

Capt. James Clark Ross, in June, 1831, found that the dip of the needle at a place whose latitude was 70° 05′ 17″ north and whose longitude was 96° 45′ 48″ west of Greenwich was 89° 59′.5. The compass needle had lost its directive force at this point entirely; when suspended by a fiber it would remain in any position in which it had been placed. This point, reached for the first time by Ross and designated the "North Magnetic Pole," is situated on Boothia Felix—named in honor of Felix Booth, who had fitted out the expedition. Owing to the method of determination which Ross had to employ and the inaccuracy of his instruments, the position found for the magnetic pole must be regarded as only approximate. To fix the point precisely would require the magnetic survey of a considerable area, and hence the expenditure of more time than Ross could afford.

A Norwegian, Mr. Roald Amundsen, is at present planning a north magnetic pole expedition, which is to set out in the spring of 1903, and is to be equipped for a stay of four years in the region of the magnetic pole. His magnetic instruments are being constructed especially for this expedition under the able superintendence of Professor Neumayer, director of the German Naval Observatory at Hamburg, and Dr. Chree, superintendent of Kew Observatory, England.

PRINCIPAL FACTS OF THE EARTH'S MAGNETISM.   75

The change in the magnetic inclination—the element principally involved in the location of the magnetic pole—along a magnetic meridian is in this region about $0'.8$ to $2'$ per kilometer, or $1'$ to $3'$ per mile. It is furthermore probable that the magnetic pole is at present moving, because of the secular change in the Earth's magnetism in a northwesterly direction at the rate of about 8–13 kilometers, or 5–8 miles, per year.

It would accordingly seem that with modern instrumental means and methods the

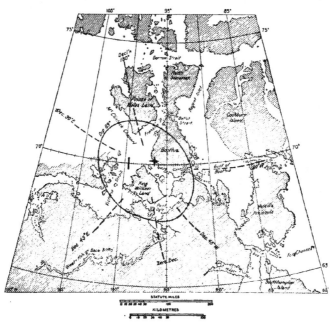

FIG. 26.—Map of region around North Magnetic Pole (Schott, 1890).

location of the magnetic pole, defined as the focus of vertical dip, and its secular motion, ought to be determinable with sufficient accuracy within the period of the expedition.

Observations of the diurnal variation of the magnetic elements, if possible, of magnetic perturbations, polar lights, and atmospheric electricity will be extremely interesting and valuable in this region.

The south magnetic pole has not as yet been reached. From Ross's observations, made in the antarctic regions while in command of the ship *Erebus*, Duperrey has deduced the position of 75° south and 138° east of Greenwich. The nearest approach to the south magnetic pole was made by Ross, February 16, 1841, in latitude 76° 20' south and longitude 165° 32' east, the dip at this place being 88° 35'. Unsurmountable physical difficulties prevented his getting any nearer.

Duperrey determined the position of the magnetic poles with the aid of his charts of the magnetic meridians for 1836. (See Fig. 23.) These meridians do not quite meet in the same point because of the irregularity in the Earth's magnetization, as already pointed out; however, the "successive intersection of each pair of contiguous meridians form a closed curve, the central points of which may be denominated magnetic poles." The poles so defined were found to be in 70° north, 98° west, and 75° south and 138° east. Gauss, in 1838, calculated that the north magnetic pole would be in 73° 35' north and 93° 39' west, and the south magnetic pole in 72° 35' south and 152° 30' east. Commander Borchgrevink, who has penetrated the farthest south thus far, claims that the position of the south magnetic pole, computed (he did not reach the pole) from his magnetic observations, gives a position agreeing more closely with that of Gauss than that of Ross. Enough has been given to show, however, that the positions of the magnetic poles are not as yet accurately known, and that, furthermore, any position determined applies only to a particular time.

MAGNETIC MOMENT.

The following figures are given to furnish some slight conception of the magnetic moment of the Earth. Suppose as the unit, a bar magnet of the hardest steel, magnetized as strongly as possible, which shall be 14 inches long, 1 inch wide, $\frac{1}{4}$ inch thick. Such a bar magnet would weigh 1 pound. According to Gauss, it would take the following number of these bar magnets placed at the Earth's center to produce the same external effect as the Earth:

$$8\ 464\ 000\ 000\ 000\ 000\ 000\ 000.$$

Or, if we assume that the Earth's magnetism is uniformly distributed throughout the Earth, then will the magnetic intensity of each cubic yard be equal to six of the 1-pound steel magnets.

To put the same fact in still another form. The radius of a soft iron sphere magnetized to saturation and concentric with the Earth, which shall have the same magnetic effect as that of the Earth, is, according to Overbeck, 243.2 kilometers, or 132.4 geographical miles, or 151 statute miles, or one-twenty-sixth of the Earth's radius.

According to Gray ("Treatise on Magnetism and Electricity," 1898): "Certain long bars of steel of comparatively high magnetizability have been found by the author to take a magnetic moment of about 780 per cubic centimeter (that is, an induction in the steel of over 10 000, about four and one-half times that taken by Gauss's bar). Consequently, the magnetic moment of a cubic centimeter of such steel is about ten times as great as that of a cubic decimeter of the Earth—that is, the mean magnetization intensity of the Earth's substance is about $\frac{1}{10000}$ of that of very highly magnetized hard steel."

Fleming says ("Terrestrial Magnetism," Vol. II, p. 58):

"Taken as a whole, the Earth is a feeble magnet. If our globe were wholly made of steel and magnetized as highly as an ordinary steel-bar magnet, the magnetic forces at its surface would be at least a hundred times as great as they are now. That might be an advantage or a very great disadvantage."

In conclusion, it may be well to point out that the actual distribution and location of the magnetic masses or systems of electric currents within the Earth's crust which cause the observed magnetic facts on the globe can not be definitely determined until observations in sufficient number and of the required accuracy have been made not only on the surface, but also at various altitudes and depths—in the upper regions and in the ocean depths. The facts measured and observed simply on the surface can be explained in an infinite number of ways.

All modern investigations would seem to lead to the conclusion that there exists both a very deep-seated magnetic field and one confined to a comparatively thin layer, and that the Earth's total magnetism results from systems of electric currents as well as from permanent and induced magnetizations.

DETERMINATION OF THE TRUE MERIDIAN AND THE MAGNETIC DECLINATION.

DETERMINATION OF THE TRUE MERIDIAN.

Such methods as can be employed with the means usually at a surveyor's disposal are described first, and then the method generally used by the magnetic observers of the Coast and Geodetic Survey is given.

SIMPLE METHODS FOR DETERMINING THE TRUE MERIDIAN BY OBSERVATIONS ON POLARIS.[a]

I.—TO DETERMINE THE TRUE MERIDIAN BY OBSERVATION ON POLARIS AT ELONGATION WITH A SURVEYOR'S TRANSIT.

1. Set a stone, or drive a wooden plug, firmly in the ground and upon the top thereof make a small distinct mark.

2. About thirty minutes before the time of the eastern or western elongation of Polaris, as given by the tables of elongation, No. VII, set up the transit firmly, with its vertical axis exactly over the mark, and carefully level the instrument.

3. Illuminate the cross hairs by the light from a bull's-eye lantern or other source, the rays being directed into the object end of the telescope by an assistant. Great care should be taken to see that the line of collimation describes a truly vertical plane.

4. Place the vertical hair upon the star, which, if it has not reached its elongation, will move to the right for eastern and to the left for western elongation.

5. As the star moves toward elongation, keep it continually covered by the vertical hair by means of the tangent screw of the vernier plate, until a point is reached where it will appear to remain on the hair for some time and then leave it in a direction contrary to its former motion, thus indicating the point of elongation.

6. At the instant the star appears to thread the vertical hair, depress the telescope to a horizontal position; about 100 yards north of the place of observation drive a wooden plug, upon which by a strongly illuminated pencil or other slender object, exactly coincident with the vertical hair, mark a point in the line of sight thus determined; then *quickly* revolve the vernier plate 180°, again place the vertical hair upon the star, and, as before, mark a point in the new direction; then the middle point between the two marks, with the point under the instrument, will define on the ground the trace of the vertical plane through Polaris at its eastern or western elongation, as the case may be.

7. By daylight lay off to the east or west, as the case may require, the proper azimuth taken from the Table No. VIII; the instrument will then define the *true meridian*, which may be permanently marked by monuments for future reference.

---

[a] In the preparation of this article use has been made of the United States Land Office Manual of Instructions, Washington, 1896.

TABLE VII.—*Local mean (astronomical) time of the culminations and elongations of Polaris in the year 1915.*

[Computed for latitude 40° north and longitude 90° or 6ʰ west of Greenwich.]

Date.	East elongation.	Upper culmination.	West elongation.	Lower culmination.
1915	h m	h m	h m	h m
January 1	0 51.7	6 46.9	12 42.1	18 44.9
January 15	23 52.5	5 51.6	11 46.8	17 49.6
February 1	22 45.3	4 44.5	10 39.7	16 42.5
February 15	21 50.1	3 49.2	9 44.4	15 47.2
March 1	20 54.8	2 54.0	8 49.2	14 52.0
March 15	19 59.6	1 58.8	7 54.0	13 56.8
April 1	18 52.7	0 51.9	6 47.1	12 49.9
April 15	17 57.7	23 52.9	5 52.0	11 54.8
May 1	16 54.8	22 50.0	4 49.2	10 52.0
May 15	15 59.9	21 55.1	3 54.2	9 57.0
June 1	14 53.3	20 48.5	2 47.6	8 50.4
June 15	13 58.5	19 53.7	1 52.8	7 55.6
July 1	12 55.9	18 51.1	0 50.2	6 53.0
July 15	12 01.1	17 56.3	23 51.5	5 58.2
August 1	10 54.5	16 49.7	22 44.9	4 51.7
August 15	9 59.8	15 55.0	21 50.2	3 56.9
September 1	8 53.2	14 48.4	20 43.6	2 50.3
September 15	7 58.3	13 53.5	19 48.7	1 55.4
October 1	6 55.5	12 50.7	18 45.9	0 52.7
October 15	6 00.6	11 55.8	17 51.0	23 53.8
November 1	4 53.7	10 48.9	16 44.1	22 46.9
November 15	3 58.6	9 53.8	15 49.0	21 51.8
December 1	2 55.6	8 50.8	14 46.0	20 48.8
December 15	2 00.4	7 55.6	13 50.8	19 53.6

A. *To refer the above tabular quantities to years other than 1915.*

        m
 For year 1908 {subtract 1.7 up to March 1
        {subtract 5.6 on and after March 1
  1909 subtract 4.4
  1910 subtract 3.2
  1911 subtract 1.8
  1912 {subtract 0.4 up to March 1
    {subtract 4.3 on and after March 1
  1913 subtract 2.9
  1914 subtract 1.5
  1916 {add 1.6 up to March 1
    {subtract 2.3 on and after March 1
  1917 subtract 0.7
  1918 add 0.9
  1919 add 2.5
  1920 {add 4.0 up to March 1
    {add 0.1 on and after March 1
  1921 add 1.6
  1922 add 3.1
  1923 add 4.5
  1924 {add 5.9 up to March 1
    {add 2.0 on and after March 1
  1925 add 3.3
  1926 add 4.6
  1927 add 5.9
  1928 {add 7.2 up to March 1
    {add 3.3 on and after March 1

TRUE MERIDIAN AND MAGNETIC DECLINATION. 81

B. *To refer to any calendar day other than the first and fifteenth of each month* SUBTRACT *the quantities below from the tabular quantity for the* PRECEDING DATE.

Day of month.		Minutes.	No. of days elapsed.
2 or 16		3.9	1
3	17	7.8	2
4	18	11.8	3
5	19	15.7	4
6	20	19.6	5
7	21	23.5	6
8	22	27.4	7
9	23	31.4	8
10	24	35.3	9
11	25	39.2	10
12	26	43.1	11
13	27	47.0	12
14	28	51.0	13
	29	54.9	14
	30	58.8	15
	31	62.7	16

C. *To refer the table to Standard time and to the civil or common method of reckoning:*

(^{a}) ADD to the tabular quantities four minutes for every degree of longitude the place is west of the Standard meridian and SUBTRACT when the place is east of the Standard meridian.

(^{b}) The astronomical day begins twelve hours after the civil day, i. e., begins at noon on the civil day of the same date, and is reckoned from 0 to 24 hours. Consequently an astronomical time less than twelve hours refers to the same civil day, whereas an astronomical time greater than twelve hours refers to the morning of the next civil day.

It will be noticed that for the tabular year two eastern elongations occur on January 14 and two western elongations on July 13. There are also two upper culminations on April 14 and two lower culminations on October 14. The lower culmination either follows or precedes the upper culmination by $11^h 58^m.0$.

D. *To refer to any other than the tabular latitude between the limits of $10°$ and $50°$ north:* ADD to the time of west elongation $0^m.10$ for every degree south of $40°$ and SUBTRACT from the time of west elongation $0^m.16$ for every degree north of $40°$. Reverse these operations for correcting times of east elongation.

E. *To refer to any other than the tabular longitude:* ADD $0^m.16$ for each $15°$ east of the ninetieth meridian and SUBTRACT $0^m.16$ for each $15°$ west of the ninetieth meridian.

TABLE VIII.—*Azimuth of Polaris when at elongation for any year between 1908 and 1928.*

Latitude.	1908	1909	1910	1911	1912	1913	1914	1915	1916	1917
	° ′	° ′	° ′	° ′	° ′	° ′	° ′	° ′	° ′	° ′
0										
10	1 12.2	1 11.8	1 11.5	1 11.2	1 11.2	1 10.9	1 10.6	1 10.3	1 10 0	1 09.6	1 09.3
11	12.4	12.1	11.8	11.4	11.1	10.8	10.5	10.2	09.9	09 6
12	12.6	12.3	12.0	11.7	11.4	11.1	10.8	10.4	10.1	09.8
13	12.9	12.6	12.3	12.0	11.7	11.4	11.0	10.7	10.4	10.1
14	13.2	12.9	12.6	12.3	12.0	11.6	11.3	11.0	10.7	10.4
15	13.6	13.2	12.9	12.6	12.3	12.0	11.6	11.3	11.0	10.7
16	13.9	13.6	13.3	13.0	12.6	12.3	12.0	11.7	11.4	11.0
17	14.3	14.0	13.7	13.4	13.0	12.7	12.4	12.0	11.7	11.4
18	14.7	14.4	14.1	13.7	13.4	13.1	12.8	12.4	12.1	11.8
19	15.2	14.8	14.5	14.2	13.8	13.5	13.2	12.8	12.5	12.2
20	15.6	15.3	15.0	14.6	14.3	14.0	13.6	13.3	13.0	12.7
21	16.1	15.8	15.4	15.1	14.8	14.5	14.1	13.8	13.5	13.1
22	16.6	16.3	16.0	15.6	15.3	15.0	14.6	14.3	14.0	13.6
23	17.2	16.9	16.5	16.2	15.9	15.5	15.2	14.8	14.5	14.2
24	17.8	17.4	17.1	16.8	16.4	16.1	15.8	15.4	15.1	14.7
25	18.4	18 1	17.7	17.4	17.0	16.7	16.4	16.0	15.7	15.3
26	19.1	18.7	18.4	18.0	17.7	17.3	17.0	16.6	16.3	16.0
27	19.8	19.4	19.1	18.7	18.4	18.0	17.7	17.3	17.0	16.6
28	20.5	20.1	19.8	19 4	19.1	18.7	18.4	18 0	17.7	17.3
29	21.3	20.9	20.5	20.2	19 8	19.5	19.1	18.8	18.4	18.1
30	22.1	21 7	21 3	21.0	20.6	20.3	19.9	19.6	19.2	18 8
31	22.9	22.5	22.2	21.8	21.5	21.1	20.7	20 4	20.0	19.7
32	23.8	23.4	23.1	22.7	22.3	22.0	21.6	21.2	20.9	20.5
33	24.7	24.3	24.0	23.6	23.3	22.9	22.5	22.1	21.8	21.4
34	25.7	25.3	25.0	24.6	24.2	23.8	23.5	23.1	22.7	22.4
35	26.8	26.4	26.0	25.6	25.2	24.9	24.5	24.1	23.7	23.3
36	27.9	27.5	27.1	26.7	26.3	25.9	25.5	25.2	24.8	24.4
37	29.0	28.6	28 2	27.8	27.4	27.0	26 7	26.3	25.9	25.3
38	30.2	29.8	29.4	29.0	28.6	28.2	27.8	27.4	27.0	26.6
39	31.4	31.0	30.6	30.2	29.8	29.4	29.0	28.6	28.2	27.8
40	32.8	32.4	32.0	31.6	31.1	30.7	30.3	29.9	29.5	29.1
41	34.2	33.8	33.4	32.9	32.5	32 1	31.7	31.3	30.9	30.4
42	35.6	35.2	34.8	34.4	34.0	33.5	33.1	32.7	32.3	31.9
43	37.2	36.8	36.3	35.9	35.5	35.0	34.6	34.2	33.8	33.4
44	38.8	38.4	37.9	37.5	37.1	36.6	36.2	35.8	35.3	34.9
45	40.5	40.1	39.6	39 2	38.7	38.3	37.8	37.4	37.0	36.6
46	42.3	41.9	41.4	41.0	40.5	40.1	39.6	39.2	38.7	38.3
47	44.2	43.7	43.3	42.8	42.4	41.9	41.5	41.0	40 6	40 1
48	46.3	45.8	45.3	44 8	44.4	43 9	43.4	43.0	42.5	42.0
49	48.4	47.9	47.4	46.9	46.4	46.0	45.5	45.0	44.5	44.1
50	1 50.6	1 50.1	1 49.6	1 49.1	1 48.6	1 48.2	1 47.7	1 47.2	1 46 7	1 46.2

TRUE MERIDIAN AND MAGNETIC DECLINATION.

TABLE VIII.—*Azimuth of Polaris when at elongation for any year between 1908 and 1928*—Concluded.

Latitude.	1918	1919	1920	1921	1922	1923	1924	1925	1926	1927	1928
°	° ′	° ′	° ′	° ′	° ′	° ′	° ′	° ′	° ′	° ′	° ′
10	1 09.0	1 08.7	1 08.4	1 08.1	1 07.8	1 07.4	1 07.2	1 06.8	1 06.5	1 06.2	1 05.9
11	09.2	08.9	08.6	08.3	08.0	07.7	07.4	07.0	06.7	06.4	06.1
12	09.5	09.2	08.9	08.6	08.2	07.9	07.6	07.3	07.0	06.7	06.4
13	09.8	09.4	09.1	08.8	08.5	08.2	07.8	07.6	07.2	06.9	06.6
14	10.0	09.7	09.4	09.1	08.8	08.5	08.2	07.8	07.5	07.2	06.9
15	10.4	10.0	09.7	09.4	09.1	08.8	08.5	08.1	07.8	07.5	07.2
16	10.7	10.4	10.1	09.8	09.4	09.1	08.8	08.5	08.2	07.8	07.5
17	11.1	10.8	10.4	10.1	09.8	09.5	09.2	08.8	08.5	08.2	07.9
18	11.5	11.1	10.7	10.5	10.2	09.8	09.5	09.2	08.9	08.6	08.2
19	11.9	11.6	11.2	10.9	10.6	10.2	09.9	09.6	09.3	09.0	08.6
20	12.3	12.0	11.7	11.4	11.0	10.7	10.4	10.0	09.7	09.4	09.1
21	12.8	12.5	12.2	11.8	11.5	11.2	10.8	10.5	10.2	09.8	09.5
22	13.3	13.0	12.6	12.3	12.0	11.6	11.3	11.0	10.6	10.3	10.0
23	13.8	13.5	13.2	12.8	12.5	12.2	11.8	11.5	11.2	10.8	10.5
24	14.4	14.1	13.7	13.4	13.0	12.7	12.4	12.0	11.7	11.4	11.0
25	15.0	14.7	14.3	14.0	13.6	13.3	13.0	12.6	12.3	11.9	11.6
26	15.6	15.3	14.9	14.7	14.2	13.9	13.6	13.2	12.9	12.5	12.2
27	16.3	15.9	15.6	15.2	14.9	14.6	14.2	13.9	13.5	13.2	12.8
28	17.0	16.6	16.3	15.9	15.6	15.2	14.9	14.6	14.2	13.8	13.5
29	17.7	17.4	17.0	16.6	16.3	16.0	15.6	15.2	14.9	14.6	14.2
30	18.5	18.1	17.8	17.4	17.0	16.7	16.4	16.0	16.6	15.3	14.9
31	19.3	18.9	18.6	18.2	17.9	17.5	17.2	16.8	16.4	16.1	15.7
32	20.1	19.8	19.4	19.1	18.7	18.3	18.0	17.6	17.2	16.9	16.5
33	21.0	20.7	20.3	19.9	19.6	19.2	18.8	18.5	18.1	17.8	17.4
34	22.0	21.6	21.2	20.9	20.5	20.1	19.8	19.4	19.0	18.6	18.3
35	23.0	22.6	22.2	21.8	21.5	21.1	20.7	20.4	20.0	19.6	19.2
36	24.0	23.6	23.3	22.9	22.5	22.1	21.7	21.4	21.0	20.6	20.2
37	25.1	24.7	24.3	24.0	23.6	23.2	22.8	22.4	22.0	21.6	21.3
38	26.2	25.9	25.5	25.1	24.7	24.3	23.9	23.5	23.2	22.8	22.4
39	27.5	27.1	26.7	26.3	25.8	25.5	25.1	24.7	24.3	23.9	23.5
40	28.7	28.3	27.9	27.5	27.1	26.7	26.3	25.9	25.5	25.1	24.7
41	30.0	29.6	29.1	28.8	28.4	28.0	27.6	27.2	26.8	26.4	26.0
42	31.5	31.0	30.6	30.2	29.8	29.4	29.0	28.6	28.2	27.8	27.3
43	32.9	32.5	32.1	31.8	31.2	30.8	30.4	30.0	29.6	29.1	28.7
44	34.5	34.1	33.6	33.2	32.8	32.4	31.9	31.5	31.1	30.6	30.2
45	36.1	35.7	35.3	34.8	34.4	34.0	33.5	33.1	32.6	32.2	31.8
46	37.8	37.4	37.0	36.5	36.1	35.6	35.2	34.8	34.3	33.9	33.4
47	39.7	39.2	38.8	38.3	37.9	37.4	37.0	36.5	36.1	35.6	35.2
48	41.6	41.1	40.7	40.2	39.8	39.3	38.8	38.4	37.9	37.4	37.0
49	43.6	43.1	42.7	42.2	41.7	41.3	40.8	40.3	39.9	39.4	38.9
50	1 45.7	1 45.3	1 44.8	1 44.3	1 43.8	1 43.4	1 42.9	1 42.4	1 41.9	1 41.4	1 41.0

The above table was computed with the mean declination of Polaris for each year. A more accurate result will be had by applying to the tabular values the following correction, which depend on the difference of the mean and the apparent place of the star. The deduced azimuth will, in general, be correct within 0′.3.

For middle of	Correction ′	For middle of	Correction ′
January	−0.5	July	+0.2
February	−0.4	August	+0.1
March	−0.3	September	−0.1
April	0.0	October	−0.4
May	+0.1	November	−0.6
June	+0.2	December	−0.8

II.—TO DETERMINE THE TRUE MERIDIAN BY OBSERVATION ON POLARIS AT ELONGATION WITH A PLUMB LINE AND PEEP SIGHT.

1. Attach the plumb line to a support situated as far above the ground as practicable, such as the limb of a tree, a piece of board nailed or otherwise fastened to a telegraph pole, a house, barn, or other building affording a clear view in a north and south direction.

The plumb bob may consist of any weighty material, such as a brick, or a piece of iron or stone, weighing 4 to 5 pounds, which will hold the plumb line straight and vertical fully as well as one of turned and finished metal.

Strongly illuminate the plumb line just below its support by a lamp or candle, care being taken to obscure the source of light from the view of the observer by an opaque screen.

For a peep sight, cut a slot about one-sixteenth of an inch wide in a thin piece of board, or nail two strips of tin, with straight edges, to a square block of wood, so arranged that they will stand vertical when the block is placed flat on its base upon a smooth horizontal rest, which will be placed at a convenient height south of the plumb line and firmly secured in an east and west direction, in such a position that when viewed through the peep sight Polaris will appear about a foot below the support of the plumb line.

The position may be determined by trial the night preceding that set for the observation.

About thirty minutes before the time of elongation, as given in the tables of elongation, bring the peep sight into the same line of sight with the plumb line and Polaris.

To reach elongation the star will move off the plumb line to the east for eastern elongation, or to the west for western elongation; therefore by moving the peep sight in the proper direction, east or west, as the case may be, keep the star on the plumb line until it appears to remain stationary, thus indicating that it has reached its point of elongation.

The peep sight will now be secured in place by a clamp or weight, and all further operations will be deferred until the next morning.

4. By daylight place a slender rod at a distance of 200 or 300 feet from the peep sight and exactly in range with it and the plumb line; carefully measure this distance.

Take from the Table VIII the azimuth of Polaris corresponding to the latitude of the station and year of observation; find the natural tangent of said azimuth and multiply it by the distance from the peep sight to the rod; the product will express the distance to be laid off from the rod *exactly at right angles* to the direction already determined (to the *west* for eastern elongation or to the *east* for western elongation) to a point which with the peep sight will define the direction of the *true meridian* with a fair degree of accuracy.

III. —TO DETERMINE THE TRUE MERIDIAN BY OBSERVING THE TRANSITS OF POLARIS AND ANOTHER STAR ACROSS THE SAME VERTICAL PLANE.

This simple method for tracing out on the ground a true north and south line, one demanding only a very slender instrumental outfit, was given in Lalande's Astronomy published more than a century ago. It was used by Andrew Ellicott in 1785 in his boundary survey work of Pennsylvania, and was again brought to notice in the present century by Dr. Charles Davies. It consists in watching for the time when Polaris and a given bright star come to the same vertical, and then after a short lapse of time, given in a table, Polaris will be found exactly on the meridian and hence can be referred to the horizon and to any meridian mark placed there.

TRUE MERIDIAN AND MAGNETIC DECLINATION.

The verticality may be ascertained by a plumb line or by the vertical thread of a transit instrument; the method demands neither a graduated circle, nor a chronometer, nor any *exact* knowledge of the local time, an ordinary watch being sufficient to measure the short tabular interval.

Early in the present century the star Alioth (ϵ Ursæ Majoris) was favorably situated for the use of the method; however, in 1850 the interval between times of verticality and of culmination already amounted to 17 minutes, and at the present time has become so large that this star is no longer suitable. Zeta (ζ) Ursæ Majoris or Delta (δ) Cassiopeiæ should now be substituted for it, both these stars being now in very favorable positions. Zeta (ζ) Ursæ Majoris, or Mizar, as it was called by the ancient Arabians, is the middle one of the three stars in the tail of the Great Bear; the small star near it is Alcor. Delta (δ) Cassiopeiæ is at the bottom of the less perfectly formed V of the letter W, as frequently imagined to unite roughly the five brightest stars of this constellation.

The diagram (Fig. 27), drawn to scale, exhibits the principal stars of the constellations Cassiopeia and Great Bear, with Delta (δ) Cassiopeia, Zeta (ζ) of the Great Bear, and *Polaris* on the meridian, represented by the straight line, Polaris being at *lower* culmination.

In employing this method the following instructions may be followed:

1. Select that one of the two stars which at the time of the year when the observation is made passes the meridian *below* Polaris. When the star passes the meridian above the pole it is too near the zenith to be of service. Delta (δ) Cassiopeiæ is on the meridian below Polaris and the pole at midnight about April 10, and is, therefore, the proper star to use at that date and for some two or three months before and after. Six months later the star Zeta (ζ) Ursæ Majoris will supply its place.

2. Using the apparatus just described under II, place the "peep sight" in the line with the plumb line and Polaris, and move it to the *west* as Polaris moves *east*, until Polaris and δ Cassiopeia, for example, *appear upon the plumb line together*, and carefully note the time by a clock or watch; then by moving the peep sight, preserve the alignment with *Polaris* and the *plumb line*

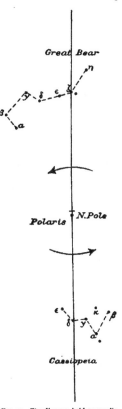

FIG. 27.—The diagram held perpendicular to the line of sight directed to the pole, with the right-hand side of the page uppermost, will represent the configuration of the constellations with Polaris near *eastern* elongation at midnight about July 11. *Inverted*, it will show Zeta (ζ) of the Great Bear and Polaris on the meridian (the former *below* and the latter *above* the pole) at midnight about October 10; and held with left-hand side uppermost, the diagram will indicate the relative situations for midnight about January 8, with Polaris near *western* elongation. The arrows indicate the direction of apparent motion.

(paying no further attention to the other star); at the expiration of the small interval of time given below the *peep sight* and *plumb line* will define the *true meridian*, which may be permanently marked for future use.

		Annual Increase.
For Zeta (ζ) Ursæ Majoris in 1912	+7.1 minutes	0.40 minute.
For Delta (δ) Cassiopeiæ in 1912	+8.2 "	0.42 "

The method given in this article for finding the true meridian can not be used with advantage at places below about 38° north latitude, on account of the haziness of the atmosphere near the horizon.

The foregoing methods for the determination of the true meridian are excellent and when available they answer the requirements of the surveyor and give results with all desirable precision. They do not require an accurate knowledge of the time, which is their principal advantage. The relative motion of the stars employed in the third method and the change in direction of motion of Polaris at elongation indicate with sufficient exactness the moment when the observation should be made. Stormy weather, a hazy atmosphere, or the presence of clouds may interfere or entirely prevent observation when the star is either at elongation or on the meridian, and both events sometimes occur in broad daylight or at an inconvenient hour of the night. Under such circumstances a simple method applicable at any time (Polaris being visible) is desirable and can often be used by the surveyor when other methods fail.

IV.—TO DETERMINE THE TRUE MERIDIAN BY MEANS OF AN OBSERVATION OF POLARIS AT ANY HOUR WHEN THE STAR IS VISIBLE, THE CORRECT LOCAL MEAN TIME BEING KNOWN.[a]

This method requires a knowledge of the local mean time within one or two minutes, as in the extreme case when Polaris is at culmination its azimuth changes 1' (arc) in 2½ minutes (time). The Standard time can usually be obtained at a telegraph office from the signals which are sent out from observatories. From this the local mean time may be derived by subtracting four minutes of time for every degree of longitude west of the Standard meridian or adding four minutes for every degree east of the Standard meridian. The local mean time may be obtained also by observations of the sun, one method being explained later.

The following table, IX, is intended to be used in connection with the American Ephemeris and Nautical Almanac. The surveyor should read carefully the chapter in that publication in which the formation and use of the Ephemeris are explained, especially the portion defining the different kinds of time.

---

[a] *cf.* Appendix No. 10, Coast and Geodetic Survey Report for 1895.

TRUE MERIDIAN AND MAGNETIC DECLINATION.

The following example explains the use of the table and the derivation of the hour angle of Polaris:

Position, latitude 36° 20′ N., longitude 80° 07′.5 or 5ʰ 20ᵐ 30ˢ W. of Greenwich.

	h.	m.	s.
Time of observation, July 10, 1908, standard (75th mer.) mean time	8	52	40 p. m.
Reduction to local time	−	20	30
Local mean time	8	32	10
Reduction to sidereal time (Table III, Amer. Ephem.)	+	01	24
Sidereal time mean noon, Greenwich, July 10, 1902	7	12	02
Correction for longitude 5ʰ 20ᵐ 30ˢ (Table III, Amer. Ephem.)	+	00	53
Local sidereal time	15	46	29
Apparent right ascension of Polaris, July 10, 1908	1	26	05
Hour angle before upper culmination	9	39	36

	°	′
Declination for which Table IX applies	88	51
Apparent declination, July 10, 1908	88	48.7
Decrease in declination	−	2.3

	°	′	″
Azimuth from Table IX (interpolated),	0	48	39
Correction for 2′.3 decrease in declination	+	1	37
Computed azimuth	0	50	16 East of north.

It is to be remembered that Polaris is east of the meridian for twelve hours before, and west of the meridian for twelve hours after, upper culmination.

Without the American Ephemeris the table may be conveniently used for obtaining the true meridian, in connection with Table VII giving the approximate mean times of culminations of Polaris, and the additional knowledge of the fact that the mean declination of Polaris is 88° 51′.1 in 1915 and increasing at the rate of about 0′.3 per year. Without the use of the Ephemeris the computation would be as follows:

	h.	m.	s.
Time of observation, July 10, 1908 standard (75th mer.) mean time	8	52	40 p. m.
Reduction to local mean time	−	20	30
Local mean time	8	32	10
Local mean time of upper culmination of Polaris (Table VII and A)	18	10	12
Mean time of observation before upper culmination	9	38	02
Reduction to sidereal time	+	01	35
Hour angle before upper culmination	9	39	37

	°	′
Declination for which Table IX applies	88	51
Mean declination 1908	88	49.0
Decrease in declination	−	2.0

	°	′	″
Azimuth from Table IX	0	48	40
Correction for 2′.0 decrease in declination	+	1	24
Computed azimuth	0	50	04 East of north.

Tables are generally given in books on surveying for reducing mean solar to sidereal time, but for this computation it is near enough to consider the correction 10ˢ an hour, as the stars gain very nearly four minutes on the Sun each day.

PRINCIPAL FACTS OF THE EARTH'S MAGNETISM.

TABLE IX.—*Azimuth of Polaris at any hour angle.*

Hour angle before or after upper culmination	Azimuth of Polaris computed for declination 88° 51'											Correction for 1' increase in declination of Polaris	
	Latitude 10°	Latitude 11°	Latitude 12°	Latitude 13°	Latitude 14°	Latitude 15°	Latitude 16°	Latitude 17°	Latitude 18°	Latitude 19°	Latitude 20°	Latitude 10°	Latitude 20°
h m	° ′ ″	° ′ ″	° ′ ″	° ′ ″	° ′ ″	° ′ ″	° ′ ″	° ′ ″	° ′ ″	° ′ ″	° ′ ″	″	″
0 15	0 04 36	0 04 37	0 04 38	0 04 39	0 04 41	0 04 42	0 04 43	0 04 45	0 04 47	0 04 48	0 04 50	−4	−4
0 30	0 09 11	0 09 13	0 09 15	0 09 17	0 09 20	0 09 23	0 09 25	0 09 29	0 09 32	0 09 36	0 09 39	−8	−8
0 45	0 13 43	0 13 46	0 13 49	0 13 53	0 13 57	0 14 01	0 14 05	0 14 10	0 14 15	0 14 20	0 14 26	−12	−13
1 00	0 18 12	0 18 16	0 18 20	0 18 25	0 18 30	0 18 35	0 18 41	0 18 47	0 18 54	0 19 01	0 19 09	−16	−17
1 15	0 22 36	0 22 41	0 22 46	0 22 52	0 22 58	0 23 05	0 23 12	0 23 20	0 23 28	0 23 37	0 23 46	−20	−21
1 30	0 26 54	0 27 00	0 27 06	0 27 13	0 27 21	0 27 29	0 27 37	0 27 46	0 27 56	0 28 07	0 28 18	−24	−25
1 45	0 31 05	0 31 12	0 31 19	0 31 27	0 31 36	0 31 45	0 31 55	0 32 06	0 32 17	0 32 29	0 32 42	−27	−29
2 00	0 35 09	0 35 16	0 35 24	0 35 33	0 35 43	0 35 53	0 36 04	0 36 16	0 36 29	0 36 43	0 36 57	−31	−32
2 15	0 39 03	0 39 11	0 39 20	0 39 30	0 39 41	0 39 52	0 40 04	0 40 18	0 40 32	0 40 47	0 41 03	−34	−36
2 30	0 42 47	0 42 56	0 43 06	0 43 17	0 43 28	0 43 41	0 43 54	0 44 09	0 44 24	0 44 40	0 44 58	−37	−39
2 45	0 46 19	0 46 29	0 46 40	0 46 52	0 47 04	0 47 18	0 47 32	0 47 48	0 48 04	0 48 22	0 48 41	−40	−42
3 00	0 49 40	0 49 51	0 50 02	0 50 15	0 50 28	0 50 42	0 50 58	0 51 15	0 51 33	0 51 52	0 52 12	−43	−45
3 15	0 52 48	0 52 59	0 53 11	0 53 24	0 53 39	0 53 54	0 54 11	0 54 28	0 54 47	0 55 07	0 55 29	−46	−48
3 30	0 55 43	0 55 54	0 56 07	0 56 21	0 56 36	0 56 52	0 57 09	0 57 28	0 57 47	0 58 09	0 58 31	−49	−51
3 45	0 58 22	0 58 34	0 58 48	0 59 02	0 59 18	0 59 35	0 59 53	1 00 12	1 00 33	1 00 55	1 01 18	−51	−53
4 00	1 00 47	1 01 00	1 01 13	1 01 28	1 01 44	1 02 02	1 02 21	1 02 41	1 03 02	1 03 25	1 03 49	−53	−55
4 15	1 02 56	1 03 09	1 03 23	1 03 38	1 03 55	1 04 13	1 04 33	1 04 53	1 05 15	1 05 39	1 06 04	−55	−57
4 30	1 04 49	1 05 02	1 05 17	1 05 33	1 05 50	1 06 08	1 06 28	1 06 49	1 07 12	1 07 36	1 08 02	−56	−59
4 45	1 06 25	1 06 39	1 06 53	1 07 10	1 07 27	1 07 46	1 08 06	1 08 28	1 08 51	1 09 15	1 09 42	−58	−61
5 00	1 07 44	1 07 58	1 08 13	1 08 29	1 08 47	1 09 06	1 09 26	1 09 48	1 10 12	1 10 37	1 11 03	−59	−62
5 15	1 08 46	1 08 59	1 09 15	1 09 31	1 09 49	1 10 08	1 10 29	1 10 51	1 11 15	1 11 40	1 12 07	−60	−62
5 30	1 09 30	1 09 43	1 09 59	1 10 15	1 10 33	1 10 52	1 11 13	1 11 35	1 11 59	1 12 25	1 12 52	−60	−63
5 45	1 09 56	1 10 09	1 10 25	1 10 41	1 10 59	1 11 18	1 11 39	1 12 01	1 12 25	1 12 51	1 13 18	−61	−63
6 00	1 10 04	1 10 17	1 10 32	1 10 49	1 11 07	1 11 26	1 11 47	1 12 09	1 12 33	1 12 58	1 13 26	−61	−64
6 15	1 09 54	1 10 07	1 10 22	1 10 39	1 11 36	1 11 15	1 11 36	1 11 58	1 12 22	1 12 47	1 13 14	−61	−63
6 30	1 09 26	1 09 39	1 09 54	1 10 10	1 10 28	1 10 46	1 11 07	1 11 29	1 11 52	1 12 17	1 12 44	−60	−63
6 45	1 08 40	1 08 58	1 09 08	1 09 23	1 09 41	1 09 59	1 10 19	1 10 41	1 11 04	1 11 29	1 11 55	−60	−62
7 00	1 07 37	1 07 50	1 08 05	1 08 19	1 08 36	1 08 54	1 09 14	1 09 35	1 09 58	1 10 22	1 10 47	−59	−61
7 15	1 06 16	1 06 29	1 06 42	1 06 57	1 07 14	1 07 32	1 07 51	1 08 11	1 08 33	1 08 57	1 09 22	−58	−60
7 30	1 04 39	1 04 51	1 05 04	1 05 19	1 05 35	1 05 52	1 06 10	1 06 30	1 06 52	1 07 15	1 07 39	−56	−58
7 45	1 02 44	1 02 56	1 03 09	1 03 23	1 03 38	1 03 55	1 04 12	1 04 32	1 04 53	1 05 15	1 05 39	−54	−56
8 00	1 00 34	1 00 45	1 00 58	1 01 11	1 01 26	1 01 42	1 01 59	1 02 18	1 02 38	1 02 59	1 03 22	−52	−55
8 15	0 58 09	0 58 19	0 58 31	0 58 44	0 58 58	0 59 13	0 59 30	0 59 47	1 00 06	1 00 27	1 00 49	−50	−53
8 30	0 55 28	0 55 38	0 55 49	0 56 02	0 56 15	0 56 29	0 56 45	0 57 02	0 57 20	0 57 39	0 58 00	−48	−50
8 45	0 52 33	0 52 43	0 52 53	0 53 05	0 53 18	0 53 31	0 53 46	0 54 02	0 54 19	0 54 38	0 54 57	−45	−48
9 00	0 49 25	0 49 33	0 49 44	0 49 55	0 50 07	0 50 19	0 50 33	0 50 48	0 51 04	0 51 21	0 51 40	−42	−45
9 15	0 46 05	0 46 13	0 46 22	0 46 32	0 46 43	0 46 55	0 47 08	0 47 21	0 47 36	0 47 52	0 48 09	−40	−42
9 30	0 42 32	0 42 40	0 42 48	0 42 57	0 43 07	0 43 18	0 43 30	0 43 43	0 43 57	0 44 11	0 44 27	−37	−38
9 45	0 38 49	0 38 56	0 39 03	0 39 12	0 39 21	0 39 31	0 39 42	0 39 53	0 40 06	0 40 19	0 40 33	−34	−35
10 00	0 34 56	0 35 02	0 35 09	0 35 16	0 35 24	0 35 33	0 35 43	0 35 53	0 36 04	0 36 16	0 36 29	−31	−3′
10 15	0 30 54	0 30 59	0 31 05	0 31 12	0 31 19	0 31 27	0 31 35	0 31 44	0 31 54	0 32 05	0 32 16	−28	−28
10 30	0 26 44	0 26 48	0 26 54	0 26 59	0 27 06	0 27 12	0 27 20	0 27 28	0 27 36	0 27 45	0 27 55	−24	−24
10 45	0 22 27	0 22 31	0 22 35	0 22 40	0 22 45	0 22 51	0 22 57	0 23 04	0 23 11	0 23 18	0 23 27	−20	−20
11 00	0 18 04	0 18 06	0 18 11	0 18 15	0 18 19	0 18 24	0 18 29	0 18 34	0 18 40	0 18 46	0 18 52	−16	−16
11 15	0 13 37	0 13 40	0 13 42	0 13 45	0 13 49	0 13 52	0 13 56	0 14 00	0 14 04	0 14 09	0 14 13	−12	−12
11 30	0 09 07	0 09 08	0 09 10	0 09 12	0 09 14	0 09 16	0 09 19	0 09 22	0 09 25	0 09 28	0 09 31	−8	−8
11 45	0 04 34	0 04 35	0 04 36	0 04 37	0 04 38	0 04 39	0 04 40	0 04 41	0 04 43	0 04 44	0 04 46	−4	−4
Elongation: Azimuth	1 10 04	1 10 18	1 10 33	1 10 49	1 11 07	1 11 26	1 11 47	1 12 09	1 12 33	1 12 58	1 13 26	−61	−64
	h m s	h m s	h m s	h m s	h m s	h m s	h m s	h m s	h m s	h m s	h m s	s	s
Hourangle	5 59 11	5 59 06	5 59 02	5 58 56	5 58 51	5 58 46	5 58 41	5 58 36	5 58 30	5 58 25	5 58 19	+2	+2

TRUE MERIDIAN AND MAGNETIC DECLINATION. 89

TABLE IX.—*Azimuth of Polaris at any hour angle*—Continued.

Hour angle before or after upper culmination	Azimuth of Polaris computed for declination 88° 51'											Correction for 1' increase in declination of Polaris	
	Latitude 20°	Latitude 21°	Latitude 22°	Latitude 23°	Latitude 24°	Latitude 25°	Latitude 26°	Latitude 27°	Latitude 28°	Latitude 29°	Latitude 30°	Latitude 20°	Latitude 30°
h m	° ′ ″	° ′ ″	° ′ ″	° ′ ″	° ′ ″	° ′ ″	° ′ ″	° ′ ″	° ′ ″	° ′ ″	° ′ ″	″	″
0 15	0 04 50	0 04 52	0 04 55	0 04 57	0 04 59	0 05 02	0 05 04	0 05 07	0 05 10	0 05 13	0 05 17	− 4	− 5
0 30	0 09 39	0 09 43	0 09 48	0 09 52	0 09 57	0 10 02	0 10 07	0 10 13	0 10 19	0 10 25	0 10 31	− 8	− 9
0 45	0 14 26	0 14 32	0 14 38	0 14 45	0 14 52	0 15 00	0 15 07	0 15 16	0 15 25	0 15 34	0 15 44	−13	−14
1 00	0 19 09	0 19 16	0 19 25	0 19 34	0 19 43	0 19 53	0 20 04	0 20 15	0 20 26	0 20 39	0 20 51	−17	−18
1 15	0 23 46	0 23 56	0 24 07	0 24 18	0 24 29	0 24 42	0 24 55	0 25 08	0 25 23	0 25 38	0 25 54	−21	−23
1 30	0 28 18	0 28 29	0 28 42	0 28 55	0 29 09	0 29 24	0 29 39	0 29 55	0 30 12	0 30 30	0 30 49	−25	−27
1 45	0 32 42	0 32 55	0 33 10	0 33 25	0 33 41	0 33 58	0 34 16	0 34 34	0 34 54	0 35 15	0 35 37	−29	−31
2 00	0 36 57	0 37 12	0 37 29	0 37 46	0 38 04	0 38 23	0 38 43	0 39 04	0 39 26	0 39 50	0 40 15	−32	−35
2 15	0 41 03	0 41 20	0 41 38	0 41 57	0 42 17	0 42 38	0 43 00	0 43 24	0 43 49	0 44 14	0 44 42	−36	−39
2 30	0 44 58	0 45 16	0 45 36	0 45 57	0 46 19	0 46 42	0 47 06	0 47 32	0 47 59	0 48 27	0 48 57	−39	−43
2 45	0 48 41	0 49 01	0 49 22	0 49 45	0 50 08	0 50 33	0 51 00	0 51 27	0 51 57	0 52 27	0 53 00	−42	−46
3 00	0 52 12	0 52 33	0 52 56	0 53 20	0 53 45	0 54 12	0 54 40	0 55 10	0 55 41	0 56 14	0 56 48	−45	−50
3 15	0 55 29	0 55 51	0 56 15	0 56 41	0 57 08	0 57 36	0 58 06	0 58 37	0 59 10	0 59 45	1 00 22	−48	−53
3 30	0 58 31	0 58 55	0 59 20	0 59 47	1 00 15	1 00 45	1 01 16	1 01 49	1 02 24	1 03 01	1 03 40	−51	−56
3 45	1 01 18	1 01 43	1 02 09	1 02 37	1 03 07	1 03 38	1 04 11	1 04 46	1 05 22	1 06 00	1 06 41	−53	−58
4 00	1 03 49	1 04 15	1 04 43	1 05 12	1 05 42	1 06 15	1 06 49	1 07 25	1 08 02	1 08 42	1 09 24	−55	−61
4 15	1 06 04	1 06 31	1 06 59	1 07 29	1 08 01	1 08 34	1 09 09	1 09 46	1 10 25	1 11 06	1 11 49	−57	−63
4 30	1 08 02	1 08 29	1 08 58	1 09 29	1 10 01	1 10 35	1 11 12	1 11 50	1 12 30	1 13 12	1 13 56	−59	−64
4 45	1 09 42	1 10 10	1 10 39	1 11 10	1 11 44	1 12 19	1 12 55	1 13 34	1 14 15	1 14 58	1 15 44	−61	−66
5 00	1 11 03	1 11 32	1 12 02	1 12 34	1 13 07	1 13 43	1 14 20	1 15 00	1 15 41	1 16 25	1 17 11	−62	−68
5 15	1 12 07	1 12 36	1 13 06	1 13 38	1 14 12	1 14 48	1 15 26	1 16 06	1 16 48	1 17 32	1 18 19	−62	−69
5 30	1 12 52	1 13 21	1 13 51	1 14 24	1 14 58	1 15 34	1 16 13	1 16 53	1 17 35	1 18 20	1 19 07	−63	−69
5 45	1 13 18	1 13 47	1 14 18	1 14 50	1 15 25	1 16 01	1 16 39	1 17 19	1 18 02	1 18 47	1 19 34	−63	−69
6 00	1 13 26	1 13 54	1 14 25	1 14 57	1 15 32	1 16 08	1 16 46	1 17 26	1 18 09	1 18 53	1 19 41	−64	−70
6 15	1 13 14	1 13 43	1 14 13	1 14 45	1 15 19	1 15 55	1 16 33	1 17 13	1 17 55	1 18 40	1 19 26	−63	−69
6 30	1 12 44	1 13 12	1 13 42	1 14 14	1 14 48	1 15 23	1 16 01	1 16 40	1 17 22	1 18 06	1 18 52	−63	−68
6 45	1 11 55	1 12 23	1 12 52	1 13 24	1 13 57	1 14 32	1 15 09	1 15 48	1 16 29	1 17 12	1 17 58	−62	−67
7 00	1 10 47	1 11 15	1 11 44	1 12 15	1 12 47	1 13 22	1 13 58	1 14 36	1 15 16	1 15 59	1 16 44	−61	−66
7 15	1 09 22	1 09 49	1 10 17	1 10 47	1 11 19	1 11 53	1 12 28	1 13 05	1 13 45	1 14 26	1 15 10	−60	−65
7 30	1 07 39	1 08 05	1 08 32	1 09 02	1 09 33	1 10 05	1 10 40	1 11 16	1 11 54	1 12 35	1 13 17	−58	−64
7 45	1 05 39	1 06 04	1 06 30	1 06 59	1 07 28	1 08 00	1 08 33	1 09 08	1 09 46	1 10 24	1 11 06	−56	−62
8 00	1 03 22	1 03 46	1 04 11	1 04 38	1 05 07	1 05 36	1 06 10	1 06 43	1 07 19	1 07 57	1 08 36	−55	−60
8 15	1 00 49	1 01 11	1 01 36	1 02 02	1 02 30	1 02 59	1 03 29	1 04 02	1 04 36	1 05 12	1 05 49	−53	−57
8 30	0 58 00	0 58 22	0 58 45	0 59 10	0 59 36	1 00 04	1 00 33	1 01 04	1 01 36	1 02 10	1 02 46	−50	−54
8 45	0 54 57	0 55 17	0 55 40	0 56 03	0 56 28	0 56 54	0 57 22	0 57 51	0 58 21	0 58 53	0 59 27	−48	−51
9 00	0 51 40	0 51 59	0 52 20	0 52 41	0 53 04	0 53 29	0 53 55	0 54 22	0 54 51	0 55 21	0 55 53	−45	−48
9 15	0 48 09	0 48 27	0 48 46	0 49 07	0 49 28	0 49 51	0 50 15	0 50 40	0 51 07	0 51 35	0 52 05	−42	−45
9 30	0 44 27	0 44 43	0 45 01	0 45 20	0 45 40	0 46 01	0 46 22	0 46 46	0 47 11	0 47 37	0 48 04	−38	−42
9 45	0 40 33	0 40 48	0 41 04	0 41 21	0 41 39	0 41 58	0 42 19	0 42 40	0 43 02	0 43 26	0 43 51	−35	−38
10 00	0 36 29	0 36 43	0 36 57	0 37 12	0 37 29	0 37 46	0 38 04	0 38 23	0 38 43	0 39 04	0 39 27	−31	−34
10 15	0 32 16	0 32 28	0 32 41	0 32 54	0 33 08	0 33 24	0 33 40	0 33 57	0 34 14	0 34 33	0 34 53	−28	−30
10 30	0 27 55	0 28 05	0 28 16	0 28 28	0 28 40	0 28 53	0 29 07	0 29 22	0 29 37	0 29 53	0 30 10	−24	−26
10 45	0 23 27	0 23 35	0 23 44	0 23 54	0 24 05	0 24 16	0 24 27	0 24 39	0 24 52	0 25 06	0 25 20	−20	−22
11 00	0 18 52	0 18 59	0 19 07	0 19 15	0 19 23	0 19 32	0 19 41	0 19 51	0 20 01	0 20 12	0 20 24	−16	−18
11 15	0 14 13	0 14 19	0 14 24	0 14 30	0 14 37	0 14 43	0 14 50	0 14 58	0 15 05	0 15 14	0 15 22	−12	−13
11 30	0 09 31	0 09 34	0 09 38	0 09 42	0 09 46	0 09 51	0 09 56	0 10 01	0 10 06	0 10 11	0 10 17	− 8	− 9
11 45	0 04 46	0 04 48	0 04 50	0 04 52	0 04 54	0 04 56	0 04 58	0 05 00	0 05 03	0 05 06	0 05 09	− 4	− 4
Elongation: Azimuth	1 13 26	1 13 55	1 14 25	1 14 58	1 15 32	1 16 08	1 16 46	1 17 27	1 18 09	1 18 54	1 19 41	−64	−69
Hour angle	h m s 5 58 19	h m s 5 58 14	h m s 5 58 08	h m s 5 58 03	h m s 5 57 57	h m s 5 57 51	h m s 5 57 45	h m s 5 57 39	h m s 5 57 33	h m s 5 57 27	h m s 5 57 21	s + 2	s + 2

90 PRINCIPAL FACTS OF THE EARTH'S MAGNETISM.

TABLE IX.—*Azimuth of Polaris at any hour angle*—Continued.

Hour angle before or after upper culmination	Azimuth of Polaris computed for declination 88° 51'											Correction for 1' increase in declination of Polaris	
	Latitude 30°	Latitude 31°	Latitude 32°	Latitude 33°	Latitude 34°	Latitude 35°	Latitude 36°	Latitude 37°	Latitude 38°	Latitude 39°	Latitude 40°	Latitude 30°	Latitude 40°
h m	° ′ ″	° ′ ″	° ′ ″	° ′ ″	° ′ ″	° ′ ″	° ′ ″	° ′ ″	° ′ ″	° ′ ″	° ′ ″	″	″
0 15	0 05 17	0 05 20	0 05 23	0 05 27	0 05 31	0 05 35	0 05 40	0 05 44	0 05 49	0 05 54	0 06 00	− 5	− 5
0 30	0 10 31	0 10 38	0 10 45	0 10 53	0 11 01	0 11 09	0 11 18	0 11 27	0 11 37	0 11 47	0 11 57	− 9	−10
0 45	0 15 44	0 15 54	0 16 04	0 16 16	0 16 27	0 16 40	0 16 53	0 17 07	0 17 21	0 17 36	0 17 52	−14	−16
1 00	0 20 51	0 21 05	0 21 19	0 21 34	0 21 50	0 22 06	0 22 24	0 22 42	0 23 01	0 23 21	0 23 42	−18	−21
1 15	0 25 54	0 26 11	0 26 28	0 26 47	0 27 06	0 27 27	0 27 48	0 28 11	0 28 34	0 28 59	0 29 26	−23	−26
1 30	0 30 49	0 31 09	0 31 30	0 31 52	0 32 15	0 32 40	0 33 05	0 33 32	0 34 00	0 34 30	0 35 01	−27	−31
1 45	0 35 37	0 36 00	0 36 24	0 36 49	0 37 16	0 37 44	0 38 14	0 38 44	0 39 17	0 39 51	0 40 27	−31	−36
2 00	0 40 15	0 40 41	0 41 08	0 41 37	0 42 07	0 42 38	0 43 12	0 43 47	0 44 23	0 45 02	0 45 42	−35	−40
2 15	0 44 42	0 45 11	0 45 41	0 46 13	0 46 46	0 47 21	0 47 58	0 48 37	0 49 18	0 50 00	0 50 45	−39	−45
2 30	0 48 57	0 49 29	0 50 02	0 50 37	0 51 13	0 51 52	0 52 32	0 53 14	0 53 59	0 54 46	0 55 35	−43	−49
2 45	0 53 00	0 53 34	0 54 10	0 54 47	0 55 27	0 56 08	0 56 52	0 57 37	0 58 25	0 59 16	1 00 09	−46	−53
3 00	0 56 48	0 57 25	0 58 03	0 58 43	0 59 25	1 00 10	1 00 56	1 01 45	1 02 37	1 03 31	1 04 28	−50	−57
3 15	1 00 22	1 01 01	1 01 41	1 02 24	1 03 08	1 03 55	1 04 45	1 05 37	1 06 31	1 07 29	1 08 29	−53	−60
3 30	1 03 40	1 04 20	1 05 03	1 05 48	1 06 35	1 07 24	1 08 16	1 09 11	1 10 08	1 11 09	1 12 12	−56	−63
3 45	1 06 41	1 07 23	1 08 08	1 08 54	1 09 44	1 10 35	1 11 30	1 12 27	1 13 27	1 14 30	1 15 36	−58	−66
4 00	1 09 24	1 10 08	1 10 54	1 11 43	1 12 34	1 13 28	1 14 24	1 15 23	1 16 26	1 17 31	1 18 40	−61	−69
4 15	1 11 49	1 12 35	1 13 23	1 14 13	1 15 06	1 16 01	1 16 59	1 18 00	1 19 05	1 20 12	1 21 23	−63	−72
4 30	1 13 56	1 14 43	1 15 32	1 16 23	1 17 18	1 18 14	1 19 14	1 20 17	1 21 23	1 22 32	1 23 45	−64	−74
4 45	1 15 44	1 16 31	1 17 21	1 18 14	1 19 09	1 20 07	1 21 08	1 22 12	1 23 20	1 24 31	1 25 45	−66	−75
5 00	1 17 11	1 18 00	1 18 51	1 19 44	1 20 40	1 21 39	1 22 41	1 23 46	1 24 55	1 26 07	1 27 23	−68	−76
5 15	1 18 19	1 19 08	1 19 59	1 20 54	1 21 50	1 22 50	1 23 53	1 24 59	1 26 08	1 27 21	1 28 38	−69	−77
5 30	1 19 07	1 19 56	1 20 48	1 21 42	1 22 40	1 23 40	1 24 43	1 25 49	1 26 59	1 28 12	1 29 30	−69	−78
5 45	1 19 34	1 20 23	1 21 15	1 22 10	1 23 07	1 24 08	1 25 11	1 26 17	1 27 27	1 28 41	1 29 58	−69	−78
6 00	1 19 41	1 20 30	1 21 22	1 22 16	1 23 13	1 24 14	1 25 17	1 26 23	1 27 33	1 28 47	1 30 04	−70	−78
6 15	1 19 26	1 20 15	1 21 07	1 22 01	1 22 58	1 23 58	1 25 01	1 26 07	1 27 17	1 28 30	1 29 46	−69	−78
6 30	1 18 52	1 19 41	1 20 32	1 21 26	1 22 22	1 23 21	1 24 24	1 25 29	1 26 38	1 27 50	1 29 06	−68	−77
6 45	1 17 58	1 18 46	1 19 36	1 20 29	1 21 25	1 22 23	1 23 24	1 24 29	1 25 37	1 26 48	1 28 03	−67	−76
7 00	1 16 44	1 17 31	1 18 20	1 19 12	1 20 06	1 21 04	1 22 04	1 23 07	1 24 14	1 25 24	1 26 37	−66	−75
7 15	1 15 10	1 15 56	1 16 44	1 17 35	1 18 28	1 19 24	1 20 23	1 21 25	1 22 30	1 23 38	1 24 50	−65	−73
7 30	1 13 17	1 14 02	1 14 49	1 15 38	1 16 30	1 17 24	1 18 21	1 19 21	1 20 25	1 21 31	1 22 41	−61	−72
7 45	1 11 06	1 11 49	1 12 34	1 13 22	1 14 12	1 15 05	1 16 00	1 16 58	1 17 59	1 19 04	1 20 11	−62	−69
8 00	1 08 36	1 09 18	1 10 01	1 10 47	1 11 36	1 12 26	1 13 20	1 14 16	1 15 14	1 16 16	1 17 21	−60	−66
8 15	1 05 49	1 06 29	1 07 11	1 07 55	1 08 41	1 09 30	1 10 21	1 11 14	1 12 11	1 13 10	1 14 12	−57	−64
8 30	1 02 46	1 03 24	1 04 04	1 04 46	1 05 29	1 06 16	1 07 04	1 07 55	1 08 49	1 09 45	1 10 44	−54	−61
8 45	0 59 27	1 00 03	1 00 40	1 01 20	1 02 01	1 02 45	1 03 31	1 04 19	1 05 10	1 06 03	1 06 59	−51	−58
9 00	0 55 53	0 56 26	0 57 02	0 57 39	0 58 18	0 58 59	0 59 43	1 00 27	1 01 14	1 02 04	1 02 57	−48	−54
9 15	0 52 05	0 52 36	0 53 09	0 53 44	0 54 20	0 54 58	0 55 38	0 56 20	0 57 04	0 57 50	0 58 39	−45	−50
9 30	0 48 04	0 48 33	0 49 03	0 49 35	0 50 08	0 50 43	0 51 20	0 51 58	0 52 39	0 53 22	0 54 07	−42	−46
9 45	0 43 51	0 44 17	0 44 44	0 45 13	0 45 44	0 46 16	0 46 49	0 47 24	0 48 01	0 48 39	0 49 20	−38	−42
10 00	0 39 27	0 39 50	0 40 15	0 40 41	0 41 08	0 41 37	0 42 07	0 42 39	0 43 12	0 43 47	0 44 24	−34	−38
10 15	0 34 53	0 35 14	0 35 35	0 35 58	0 36 22	0 36 48	0 37 14	0 37 42	0 38 11	0 38 43	0 39 15	−30	−34
10 30	0 30 10	0 30 28	0 30 47	0 31 07	0 31 28	0 31 50	0 32 12	0 32 37	0 33 02	0 33 29	0 33 57	−26	−29
10 45	0 25 20	0 25 35	0 25 51	0 26 08	0 26 25	0 26 43	0 27 03	0 27 23	0 27 44	0 28 07	0 28 30	−22	−24
11 00	0 20 24	0 20 36	0 20 49	0 21 02	0 21 16	0 21 31	0 21 46	0 22 03	0 22 20	0 22 38	0 22 57	−18	−20
11 15	0 15 22	0 15 31	0 15 41	0 15 50	0 16 00	0 16 13	0 16 24	0 16 37	0 16 50	0 17 03	0 17 17	−13	−15
11 30	0 10 17	0 10 23	0 10 29	0 10 36	0 10 43	0 10 51	0 10 58	0 11 07	0 11 15	0 11 24	0 11 34	− 9	−10
11 45	0 05 09	0 05 12	0 05 15	0 05 19	0 05 22	0 05 26	0 05 30	0 05 34	0 05 38	0 05 43	0 05 47	− 4	− 5
Elongation: Azimuth.	1 19 41	1 20 30	1 21 22	1 22 17	1 23 14	1 24 14	1 25 18	1 26 24	1 27 34	1 28 48	1 30 05	−69	−78
Hour angle	h m s 5 57 21	h m s 5 57 14	h m s 5 57 08	h m s 5 57 01	h m s 5 56 54	h m s 5 56 47	h m s 5 56 39	h m s 5 56 32	h m s 5 56 24	h m s 5 56 16	h m s 5 56 08	+ 2	+ 3

TRUE MERIDIAN AND MAGNETIC DECLINATION.

TABLE IX.—*Azimuth of Polaris at any hour angle*—Continued.

Hour angle before or after upper culmination	Azimuth of Polaris computed for declination 88° 51'											Correction for 1' increase in declination of Polaris	
	Latitude 40°	Latitude 41°	Latitude 42°	Latitude 43°	Latitude 44°	Latitude 45°	Latitude 46°	Latitude 47°	Latitude 48°	Latitude 49°	Latitude 50°	Latitude 40°	Latitude 50°
h m	° ′ ″	° ′ ″	° ′ ″	° ′ ″	° ′ ″	° ′ ″	° ′ ″	° ′ ″	° ′ ″	° ′ ″	° ′ ″	″	″
0 15	0 06 00	0 06 05	0 06 11	0 06 17	0 06 24	0 06 31	0 06 38	0 06 46	0 06 54	0 07 03	0 07 12	− 5	− 6
0 30	0 11 57	0 12 09	0 12 21	0 12 33	0 12 46	0 13 00	0 13 14	0 13 30	0 13 46	0 14 03	0 14 21	−10	−13
0 45	0 17 52	0 18 09	0 18 27	0 18 45	0 19 05	0 19 25	0 19 47	0 20 10	0 20 34	0 21 00	0 21 27	−16	−19
1 00	0 23 43	0 24 04	0 24 28	0 24 52	0 25 18	0 25 45	0 26 14	0 26 45	0 27 17	0 27 51	0 28 27	−21	−25
1 15	0 29 26	0 29 53	0 30 22	0 30 53	0 31 25	0 31 59	0 32 34	0 33 12	0 33 52	0 34 34	0 35 18	−26	−32
1 30	0 35 01	0 35 34	0 36 08	0 36 45	0 37 23	0 38 03	0 38 46	0 39 30	0 40 18	0 41 08	0 42 01	−31	−38
1 45	0 40 27	0 41 05	0 41 45	0 42 27	0 43 11	0 43 57	0 44 46	0 45 38	0 46 33	0 47 30	0 48 31	−36	−43
2 00	0 45 42	0 46 25	0 47 10	0 47 57	0 48 47	0 49 39	0 50 35	0 51 33	0 52 35	0 53 40	0 54 49	−40	−49
2 15	0 50 45	0 51 33	0 52 22	0 53 15	0 54 10	0 55 08	0 56 09	0 57 14	0 58 22	0 59 35	1 00 51	−45	−54
2 30	0 55 35	0 56 26	0 57 21	0 58 18	0 59 18	1 00 22	1 01 29	1 02 40	1 03 54	1 05 13	1 06 37	−49	−59
2 45	1 00 09	1 01 05	1 02 04	1 03 06	1 04 11	1 05 20	1 06 32	1 07 48	1 09 09	1 10 34	1 12 04	−53	−64
3 00	1 04 28	1 05 28	1 06 30	1 07 36	1 08 46	1 10 00	1 11 17	1 12 39	1 14 05	1 15 36	1 17 12	−57	−68
3 15	1 08 29	1 09 32	1 10 39	1 11 49	1 13 03	1 14 21	1 15 43	1 17 10	1 18 41	1 20 18	1 22 00	−60	−72
3 30	1 12 12	1 13 19	1 14 29	1 15 43	1 17 00	1 18 22	1 19 49	1 21 20	1 22 56	1 24 37	1 26 25	−63	−76
3 45	1 15 36	1 16 46	1 17 59	1 19 16	1 20 37	1 22 03	1 23 33	1 25 08	1 26 49	1 28 35	1 30 27	−66	−80
4 00	1 18 40	1 19 52	1 21 08	1 22 28	1 23 53	1 25 21	1 26 54	1 28 34	1 30 18	1 32 08	1 34 05	−69	−83
4 15	1 21 23	1 22 38	1 23 56	1 25 19	1 26 46	1 28 17	1 29 54	1 31 33	1 33 24	1 35 17	1 37 17	−72	−86
4 30	1 23 45	1 25 02	1 26 22	1 27 47	1 29 16	1 30 50	1 32 30	1 34 14	1 36 06	1 38 01	1 40 04	−74	−88
4 45	1 25 45	1 27 03	1 28 26	1 29 52	1 31 23	1 32 59	1 34 41	1 36 28	1 38 20	1 40 19	1 42 25	−75	−90
5 00	1 27 23	1 28 42	1 30 06	1 31 34	1 33 06	1 34 44	1 36 27	1 38 16	1 40 10	1 42 11	1 44 19	−76	−91
5 15	1 28 38	1 29 58	1 31 23	1 32 52	1 34 25	1 36 04	1 37 48	1 39 38	1 41 34	1 43 36	1 45 45	−77	−92
5 30	1 29 30	1 30 50	1 32 16	1 33 45	1 35 20	1 36 59	1 38 44	1 40 34	1 42 31	1 44 34	1 46 44	−78	−93
5 45	1 29 58	1 31 20	1 32 45	1 34 15	1 35 50	1 37 29	1 39 14	1 41 05	1 43 02	1 45 05	1 47 16	−78	−94
6 00	1 30 04	1 31 25	1 32 50	1 34 20	1 35 55	1 37 34	1 39 19	1 41 09	1 43 06	1 45 09	1 47 19	−78	−93
6 15	1 29 46	1 31 07	1 32 32	1 34 01	1 35 35	1 37 14	1 38 58	1 40 48	1 42 44	1 44 46	1 46 56	−78	−93
6 30	1 29 06	1 30 26	1 31 50	1 33 18	1 34 51	1 36 29	1 38 12	1 40 01	1 41 56	1 43 57	1 46 04	−77	−92
6 45	1 28 03	1 29 20	1 30 44	1 32 11	1 33 43	1 35 19	1 37 01	1 38 48	1 40 41	1 42 40	1 44 46	−76	−91
7 00	1 26 37	1 27 54	1 29 15	1 30 41	1 32 11	1 33 45	1 35 25	1 37 10	1 39 01	1 40 58	1 43 02	−75	−89
7 15	1 24 50	1 26 05	1 27 24	1 28 48	1 30 16	1 31 48	1 33 25	1 35 08	1 36 56	1 38 51	1 40 51	−73	−87
7 30	1 22 41	1 23 54	1 25 11	1 26 32	1 27 58	1 29 27	1 31 02	1 32 42	1 34 27	1 36 18	1 38 16	−72	−85
7 45	1 20 11	1 21 22	1 22 36	1 23 55	1 25 17	1 26 44	1 28 16	1 29 52	1 31 34	1 33 22	1 35 15	−69	−82
8 00	1 17 21	1 18 29	1 19 41	1 20 57	1 22 16	1 23 40	1 25 08	1 26 41	1 28 19	1 30 02	1 31 51	−66	−79
8 15	1 14 12	1 15 17	1 16 26	1 17 38	1 18 54	1 20 14	1 21 39	1 23 07	1 24 41	1 26 20	1 28 05	−64	−76
8 30	1 10 44	1 11 46	1 12 52	1 14 00	1 15 12	1 16 29	1 17 49	1 19 14	1 20 43	1 22 17	1 23 56	−61	−72
8 45	1 06 59	1 07 57	1 08 59	1 10 04	1 11 12	1 12 23	1 13 40	1 15 00	1 16 24	1 17 53	1 19 27	−58	−68
9 00	1 02 57	1 03 52	1 04 50	1 05 50	1 06 55	1 08 02	1 09 13	1 10 28	1 11 47	1 13 10	1 14 38	−54	−64
9 15	0 58 39	0 59 30	1 00 24	1 01 21	1 02 20	1 03 23	1 04 29	1 05 39	1 06 52	1 08 10	1 09 32	−50	−59
9 30	0 54 07	0 54 54	0 55 44	0 56 36	0 57 31	0 58 28	0 59 29	1 00 34	1 01 41	1 02 53	1 04 08	−46	−55
9 45	0 49 21	0 50 04	0 50 49	0 51 37	0 52 27	0 53 20	0 54 15	0 55 13	0 56 15	0 57 20	0 58 29	−42	−50
10 00	0 44 24	0 45 02	0 45 43	0 46 25	0 47 10	0 47 58	0 48 47	0 49 40	0 50 35	0 51 34	0 52 35	−38	−45
10 15	0 39 15	0 39 49	0 40 25	0 41 03	0 41 42	0 42 24	0 43 08	0 43 54	0 44 43	0 45 35	0 46 29	−34	−40
10 30	0 33 57	0 34 26	0 34 57	0 35 30	0 36 04	0 36 40	0 37 18	0 37 58	0 38 40	0 39 24	0 40 12	−29	−34
10 45	0 28 30	0 28 55	0 29 21	0 29 48	0 30 17	0 30 47	0 31 19	0 31 53	0 32 28	0 33 05	0 33 45	−24	−29
11 00	0 22 57	0 23 16	0 23 37	0 23 59	0 24 22	0 24 47	0 25 12	0 25 39	0 26 08	0 26 38	0 27 09	−20	−23
11 15	0 17 17	0 17 32	0 17 48	0 18 05	0 18 22	0 18 40	0 19 00	0 19 20	0 19 42	0 20 04	0 20 28	−15	−18
11 30	0 11 34	0 11 44	0 11 54	0 12 06	0 12 17	0 12 29	0 12 42	0 12 56	0 13 10	0 13 25	0 13 43	−10	−12
11 45	0 05 47	0 05 52	0 05 58	0 06 05	0 06 09	0 06 15	0 06 23	0 06 32	0 06 38	0 06 44	0 06 51	− 5	− 6
Elongation: Azimuth	1 30 05	1 31 26	1 32 51	1 34 21	1 35 56	1 37 35	1 39 20	1 41 11	1 43 08	1 45 11	1 47 21	−78	−93
	h m s	h m s	h m s	h m s	h m s	h m s	h m s	h m s	h m s	h m s	h m s	s	s
Hour angle	5 56 08	5 56 00	5 55 51	5 55 43	5 55 33	5 55 24	5 55 14	5 55 04	5 54 53	5 54 42	5 54 31	+ 3	+ 3

DETERMINATION OF THE TRUE MERIDIAN AND LOCAL MEAN TIME BY MEANS OF OBSERVATIONS ON THE SUN.

The following method is the one usually employed to determine the true meridian in connection with the magnetic observations of the Coast and Geodetic Survey. It involves more computing than those already described, but is more convenient in that it is available for use during daylight when the magnetic observations are in progress. In connection with the time signals sent out by telegraph from observatories it furnishes the means also of determining approximately the longitude of the place of observation. It requires a theodolite with graduated vertical circle and a prismatic eyepiece for observing the Sun, and a well-regulated timepiece. The observations at a place usually consist of four independent sets of observations, two in the morning and two in the afternoon, each set consisting of four pointings on the Sun and two pointings on a reference mark symmetrically arranged as in the following example. For each pointing on the Sun the time is noted, and both horizontal and vertical circles are read. Observations are made from two to four hours from noon, and at nearly the same altitudes morning and afternoon. The reference mark should be a well-defined object nearly in the horizon and at least 100 yards distant.

The instrument (see Figure 28) used in the following observations has a glass diaphragm on which is ruled one horizontal and one vertical line. The symbols in the first column indicate the limbs of the Sun which were brought tangent to the lines of the diaphragm at the recorded time. The vertical circle is so graduated that it gives altitudes in one position and zenith distances in the other. The readings in the latter case have been subtracted from 90° when filling in the last column. The verniers allow readings on the horizontal and on the vertical circle to be made to minutes, half minutes being estimated.

A. M. observations of Sun for azimuth and time.

Station, Paducah, Ky.
Theodolite of Mag'r No. 19.
Chronometer, Bond No. 175.

Date, Tuesday, July 2, 1901.
Observer, W. W.
Temperature, 32°. 2.

Sun's limb	V. C.	Chronometer time	Horizontal circle			Vertical circle		
			A	B	Mean	A	B	Mean
			° ′	′	° ′			
	R	Mark	352 39.5	37.5	352 38.5			
	L		172 37.0	36.0	36.5			
		h m s				° ′	′	° ′
◐	L	9 35 15	291 41.0	39.5	111 40.2	44 17.0	18.0	44 17.50
◐	L	36 10	291 49.5	48.5	111 49.0	44 29.0	29.0	44 29.00
◑	R	37 40	112 47.0	44.0	112 45.5	44 48.0	49.5	45 11.25
◑	R	38 47	112 57.0	56.0	112 56.5	44 35.0	36.5	45 24.25
Means		9 36 58.0			112 17.8			44 50.50
						Refr. and Par.		−0.78
◐	R	9 39 46	113 07.0	05.5	113 06.2	44 23.0	25.0	45 36.00
◐	R	40 49	113 16.0	18.0	113 17.0	44 10.5	12.0	45 48.75
◑	L	42 20	292 50.0	48.0	112 49.0	45 41.5	42.0	45 41.75
◑	L	43 24	292 60.0	58.5	112 59.2	45 54.0	55.0	45 54.50
Means		9 41 34.8			113 02.8			45 45.25
						Refr. and Par.		−.76
	L	Mark	172 38.0	37.5	352 37.8			
	R		352 40.0	38.0	39.0			
					352 37.9			

TRUE MERIDIAN AND MAGNETIC DECLINATION.

P. M. observations of Sun for azimuth and time.

Station, Paducah, Ky.
Theodolite of Mag'r No. 19.
Chronometer, Bond No. 175.

Date, Tuesday, July 2, 1901.
Observer, W. W.
Temperature, 36.°8.

Sun's limb	V. C.	Chronometer time	Horizontal circle			Vertical circle		
			A	B	Mean	A	B	Mean
	N L	Mark	° ′ 112 20.5 292 20.5	′ 19.0 19.0	° ′ 112 19.8 19.8			
☉☉☉☉	L L R R	h m s 4 21 28 22 26 23 45 25 04	226 01.0 226 10.0 45 32.0 45 45.0	01.0 11.0 35.0 48.0	46 01.0 46 10.5 45 33.5 45 46.5	° ′ 44 13.5 44 02.0 45 49.0 46 03.0	′ 14.0 03.0 47.0 04.5	° ′ 44 13.75 44 02.50 44 12.00 43 56.25
Means		4 23 10.8			45 52.9	Refr. and Par.		44 06.12 −.79
☉☉☉☉	R R L L	4 34 24 35 36 37 12 38 19	47 17.0 47 25.0 228 28.0 228 38.0	14.5 27.0 26.0 37.5	47 15.8 47 26.0 48 27.0 48 37.8	47 53.0 48 07.5 41 07.0 40 53.0	56.0 10.0 08.0 53.5	42 05.50 41 51.25 41 07.50 40 53.25
Means		4 36 22.8			47 56.6	Refr. and Par.		41 29.38 −.89
	L R	Mark	292 21.0 112 21.0	19.5 19.5	112 20.2 20.2 112 20.0			

For computing these observations one requires a five-place table of logarithms of trigonometric functions and the American Ephemeris, or U. S. Hydrographic Office Publication No. 118, which gives the Sun's apparent declination and the equation of time. For correcting the observed altitude of the Sun for parallax and refraction, the following convenient table has been prepared, giving the combined correction for different altitudes and temperatures, to be subtracted from the observed altitude:

TABLE X.—*Correction to observed altitude of the Sun for refraction and parallax.*

App't Alt.	Temperature										App't Alt.
	−10° C.	−5° C.	0° C.	+5° C.	+10° C.	+15° C.	+20° C.	+25° C.	+30° C.	+35° C.	
°	′	′	′	′	′	′	′	′	′	′	°
10	5.52	5.42	5.30	5.20	5.10	5.00	4.92	4.83	4.75	4.67	10
11	5.02	4.92	4.82	4.73	4.63	4.55	4.47	4.38	4.32	4.23	11
12	4.60	4.50	4.42	4.33	4.25	4.17	4.10	4.03	3.97	3.88	12
13	4.23	4.15	4.07	4.00	3.92	3.85	3.78	3.72	3.65	3.58	13
14	3.92	3.83	3.77	3.70	3.62	3.55	3.50	3.45	3.37	3.32	14
15	3.65	3.58	3.50	3.43	3.37	3.32	3.25	3.20	3.13	3.08	15
16	3.43	3.35	3.30	3.23	3.17	3.12	3.07	3.00	2.95	2.90	16
17	3.22	3.15	3.10	3.03	2.98	2.92	2.88	2.82	2.77	2.72	17
18	3.02	2.95	2.90	2.85	2.80	2.75	2.70	2.65	2.60	2.55	18
19	2.83	2.78	2.73	2.68	2.63	2.58	2.53	2.48	2.43	2.40	19

TABLE X.—*Correction to observed altitude of the Sun for refraction and parallax*—Concl'd.

App't Alt.	Temperature.									App't Alt.
	−10° C.	−5 °C.	0° C.	+5° C.	+10° C.	+15° C.	+20° C.	+25° C.	+30° C.	+35° C.	
°	′	′	′	′	′	′	′	′	′	′	°
20	2.68	2.63	2.58	2.53	2.48	2.43	2.38	2.33	2.30	2.27	20
21	2.53	2.48	2.43	2.38	2.35	2.30	2.27	2.22	2.17	2.13	21
22	2.38	2.35	2.30	2.25	2.22	2.18	2.13	2.08	2.05	2.02	22
23	2.28	2.25	2.20	2.15	2.12	2.08	2.03	1.98	1.95	1.93	23
24	2.17	2.13	2.08	2.05	2.02	1.98	1.93	1.88	1.87	1.83	24
25	2.07	2.03	1.98	1.95	1.92	1.88	1.83	1.80	1.77	1.75	25
26	1.99	1.95	1.90	1.87	1.83	1.80	1.75	1.72	1.70	1.67	26
27	1.88	1.85	1.82	1.78	1.75	1.72	1.68	1.63	1.62	1.60	27
28	1.80	1.77	1.72	1.70	1.67	1.63	1.60	1.57	1.53	1.52	28
29	1.72	1.68	1.65	1.63	1.60	1.57	1.53	1.50	1.47	1.46	29
30	1.65	1.62	1.58	1.57	1.53	1.50	1.47	1.45	1.42	1.40	30
32	1.53	1.50	1.47	1.45	1.42	1.38	1.35	1.33	1.30	1.28	32
34	1.41	1.37	1.35	1.32	1.30	1.27	1.23	1.23	1.20	1.18	34
36	1.30	1.27	1.25	1.22	1.20	1.18	1.15	1.13	1.10	1.08	36
38	1.20	1.18	1.15	1.13	1.12	1.10	1.07	1.05	1.02	1.02	38
40	1.11	1.10	1.07	1.05	1.03	1.02	0.98	0.97	0.95	0.93	40
42	1.03	1.00	0.98	0.97	0.95	0.93	0.90	0.88	0.87	0.87	42
44	0.96	0.93	0.92	0.90	0.88	0.87	0.85	0.83	0.82	0.80	44
46	0.89	0.88	0.87	0.85	0.83	0.82	0.80	0.78	0.77	0.75	46
48	0.83	0.82	0.80	0.78	0.77	0.75	0.73	0.72	0.70	0.68	48
50	0.77	0.75	0.73	0.72	0.70	0.68	0.67	0.67	0.65	0.63	50
55	0.63	0.62	0.60	0.60	0.58	0.57	0.57	0.55	0.53	0.52	55
60	0.52	0.52	0.50	0.50	0.48	0.47	0.47	0.45	0.45	0.43	60
65	0.42	0.40	0.40	0.40	0.38	0.38	0.37	0.37	0.35	0.33	65
70	0.32	0.32	0.32	0.30	0.30	0.30	0.28	0.28	0.28	0.27	70
75	0.23	0.23	0.23	0.22	0.22	0.22	0.20	0.20	0.20	0.18	75
80	0.15	0.15	0.13	0.13	0.13	0.13	0.13	0.12	0.12	0.12	80
85	0.07	0.07	0.07	0.07	0.07	0.07	0.07	0.05	0.05	0.05	85
90	0.00	0.00	0.00	0.00	0.00	0.00	0.00	0.00	0.00	0.00	90

The formulæ used in computing the azimuth and local mean time from observations of the Sun made in the manner just described are the following:

$$\operatorname{ctn}^2 \tfrac{1}{2} A = \frac{\sin(s-\varphi)\sin(s-h)}{\cos s \cos(s-p)}$$

$$= \sec s \sec(s-p) \sin(s-h) \sin(s-\varphi)$$

$$\tan \tfrac{1}{2} t = \frac{\sin(s-h)\sec(s-p)}{\operatorname{ctn} \tfrac{1}{2} A}$$

A = azimuth of Sun, east of south in the morning, west of south in the afternoon.
φ = latitude of the place.
h = altitude of the Sun corrected for refraction and parallax in altitude.
p = Polar distance of the Sun, at the time of observation, taken from the American Ephemeris, or H. O. Publication No. 118.
$s = \tfrac{1}{2}(h+\varphi+p)$.
t = The hour angle of the Sun or apparent time of observation expressed in arc.

By combining the azimuth of the Sun with the angle between the Sun and mark, the azimuth of the mark may be obtained. This is counted from 0° to 360° from south

TRUE MERIDIAN AND MAGNETIC DECLINATION.

around by west. When the azimuth of the mark is known the true meridian may be laid off at any time by turning off the proper angle.

The apparent time of observation must be corrected for equation of time (taken from the Ephemeris), in order to obtain the local mean time. The following is a convenient form of computation:

Specimen computation of azimuth and longitude.

Date	Tuesday, July 2, 1901			
	° ′	° ′	° ′	° ′
h	44 49.7	45 44.5	44 05.3	41 28.5
ϕ	37 03.6	37 03.6	37 03.6	37 03.6
p	66 55.5	66 55.5	66 56.7	66 56.8
$2s$	148 48.8	149 43.6	148 05.6	145 28.9
s	74 24.4	74 51.8	74 02.8	72 44.4
$s-p$	7 28.9	7 56.3	7 06.1	5 47.6
$s-h$	29 34.7	29 07.3	29 57.5	31 15.9
$s-\phi$	37 20.8	37 48.2	36 59.2	35 40.8
log sec s	0.57056	0.58316	0.56090	0.52767
" sec $(s-p)$	0.00371	0.00418	0.00334	0.00222
" sin $(s-h)$	9.69339	9.68723	9.69842	9.71516
" sin $(s-\phi)$	9.78293	9.78743	9.77933	9.76586
" ctn² ½ A	0.05059	0.06200	0.04199	0.01091
" ctn ½ A	0.02530	0.03100	0.02100	0.00546
	° ′	° ′	° ′	° ′
A from South	86 39.8	85 54.8	87 13.8	89 16.8
Circle reads	112 17.8	113 02.8	45 52.9	47 56.6
S. Mer. "	198 57.6	198 57.6	318 39.1	318 39.8
Mark "	352 37.9	352 37.9	112 20.0	112 20.0
Az. of Mark	153 40.3	153 40.3	153 40.9	153 40.2
Mean	153 40.4			
log sec $(s-p)$ sin $(s-h)$	9.69710	9.69141	9.70176	9.71738
" tan ½ t	9.67180	9.66041	9.68076	9.71192
t in arc	50° 19′ 00″	49° 10′ 12″	51° 13′ 57″	54° 30′ 32″
	h m s	h m s	h m s	h m s
t	−3 21 16.0	−3 16 40.8	3 24 55.8	3 38 02.1
E	3 40.2	3 40.2	3 43.4	3 43.5
Local M. T.	8 42 24.2	8 46 59.4	3 28 39.2	3 41 45.6
Chron. time	9 36 58.0	9 41 34.8	4 23 10.8	4 36 22.8
Δt on L. M. T.	− 54 33.8	− 54 35.4	− 54 31.6	− 54 37.2
Δt on 75 M. T.	− 6.8	− 6.8	− 6.9	− 6.9
$\Delta \lambda$	54 27.0	54 28.6	54 24.7	54 30.3
Mean	54 27.6	=13° 36′.9	$\lambda=$	88° 36′.9

DETERMINATION OF THE MAGNETIC DECLINATION.

A.—With an Ordinary Compass or Surveyor's Transit.

When the surveyor determines the value of the magnetic declination himself it would be well for him to make the observations on several days, if possible, and probably the best time of the day would be toward evening, about 5 or 6 o'clock. At this time the declination reaches, approximately, its mean value for the day and is almost stationary. (See Tables III and IV.) Between 10 and 11 a. m. the declination also reaches its mean value, but it changes more rapidly than at 5 or 6 o'clock in the evening. The observations on any one day should extend at least over one-half of an hour, preferably an hour, and the readings should be taken every ten minutes. Before each reading of the needle it would be well to tap[a] the glass lightly with the finger or a pencil, so as to slightly disturb the needle from the position of rest it may have assumed. The accurate time should be noted opposite each reading and a note entered in the record book as to date, the weather, and the kind of time the observer's watch was keeping. A brief description of station and of method employed in determining the meridian line and declination should be added to the record.

Of course the instrument should be put in good adjustment and in first-class condition in every respect beforehand, and the readings should be made in such a manner as to eliminate any outstanding error of eccentricity, whether due to pivot of needle not being exactly over center of graduated circle, or to the needle being bent or the line of sight not passing through the zero points of the circle. In addition, it is very desirable that the surveyor should have some knowledge as to any constant error his instrument may be subject to, due to whatever cause, e. g., imperfect elimination of errors of adjustment or to the fact that the metal of the various parts may not be entirely free of traces of iron, or that the magnetic axis of the needle may not coincide with its geometric axis, etc. He can determine his constant error by making observations at one of the magnetic survey stations, or, better still, compare his instrument with a standard magnetometer or transit when opportunity affords. It would not be amiss to determine the compass correction before and after the determination of the magnetic declination.[b]

If these precautions are taken it is possible to determine the magnetic declination with a good transit with all needful accuracy. With special care results that will compare very favorably with those obtained by more elaborate instruments can be reached.

B.—With a Magnetometer.[c]

Special instruments, termed magnetometers, have been devised for determining accurately and expeditiously the magnetic declination and the intensity of the magnetic force. The essential feature of all is a cylindrical (or octagonal) bar magnet,

---

[a] Great care must be taken not to produce static electric charges by rubbing the glass plate in any manner. Remarkable deflections of the needle can thus be produced.

[b] Surveyor's compasses have been found to differ at times as much as $\frac{1}{2}°$ to $1°$ from the readings with the Coast and Geodetic Survey magnetometers.

[c] For a further description of methods and instruments, the reader is referred to the special paper giving directions for measurements in terrestrial magnetism; Appendix 8, Coast and Geodetic Survey Report for 1881; a new edition is now in preparation. The present purpose is simply to give a specimen of the general method employed without going into great detail.

FIG. 28.—COAST AND GEODETIC SURVEY MAGNETOMETER.

suspended by two or three silk fibers and capable of being inverted in its stirrup, the magnet taking the place of the magnetic needle in the ordinary surveyor's compass.

The fiber suspension avoids the friction incident to the use of a pivot, and by making part of the observations with magnet erect and part with magnet inverted it is possible to eliminate the error arising from lack of coincidence of the magnetic and geometric axes.

The form of magnetometer which has been in general use by the Coast and Geodetic Survey is shown in Fig. 28. It is really a combination of magnetometer and theodolite. The latter, shown at the right of the figure, can be quickly mounted in place of the magnetometer and is used for determining the true meridian, as explained in the preceding pages, and the longitude and latitude. The magnetometer is shown in position for observing declination, except that one side of the magnet box has been removed to show the manner of suspending the magnet. The magnet used in this instrument is an octagonal hollow steel bar about 3 inches long and half an inch in diameter. The south end is closed by a plane glass on which has been etched a graduated scale divided into two minute spaces (0.1 of a division being estimated), while in the north end is a collimating lens so arranged that when the small reading telescope is focused on a distant object the graduated scale will be in focus also. The magnet is supported in a brass stirrup consisting of three rings joined to a shank about an inch long. In the upper end of this shank is an eye to which one end of the silk fibers is fastened. The other end of the fibers is fastened to a suitable arrangement at the top of the glass suspension tube, by means of which the magnet may be raised to the level of the observing telescope. Light to illuminate the scale of the magnet is admitted through a hole in the south end of the magnet box with the aid of an adjustable mirror, if necessary. The north end of the magnet box is connected with the object end of the reading telescope by means of a hood of dark cloth, so that no glass comes between the objective and the magnet and air currents are excluded by the hood. The wooden sides of the magnet box may be removed to permit the necessary manipulation of the magnet. The long shank of the stirrup obviates the necessity of having a weight on the south end of the magnet to counterbalance the dip of the north end. When not in use the magnet is kept in a wooden case with its north end down, so that the effect of the Earth's magnetism may be rather to increase than decrease the strength of the magnet and thus assist in overcoming the gradual loss of the magnetic strength with time; the stirrup is fastened to a hook in the top of the magnet box to prevent the fibers from becoming twisted or broken.

The determination of the magnetic declination consists of two parts; first, the determination of the true meridian as described in the preceding pages, and second, the determination of the magnetic meridian. The method of performing the second operation with the above-described instrument is as follows: Mount the magnetometer, which is supposed to have been put in good adjustment, and level carefully by means of the striding level. Place the magnetometer so that sides of box will be parallel approximately to the magnetic meridian. Suspend the torsion weight (a solid brass cylinder of the same weight as the magnet) and replace, if need be, the wooden sides of the magnet box with others of glass. Watch the vibration of this weight and turn the torsion head at the top of the suspension tube until the torsion weight hangs parallel to the sides of the magnet box. The suspension fibers are then free from twist. Remove the torsion weight, open the glass window at the south end of the magnet box, and point upon the object selected as a reference mark in the observations to determine the true meridian. Read the two verniers of the horizontal circle and enter readings in the record. Then close the window again, turn the instrument until the telescope points approximately south (magnetic), suspend the magnet with its scale erect, raise it to the level of the reading telescope, and put back the wooden sides of the magnet box. Next turn the instrument until the division of the scale nearest to the reading of the magnetic axis coincides approximately with the vertical line in the diaphragm of the reading telescope, clamp the horizontal circle, and read both verniers. Check the vibrations of the magnet by means of a bit of steel or iron until the magnet swings over 1–2 divisions of the scale, and take the extreme readings of the scale of the swinging magnet several times at intervals of one minute, recording also the time. The magnet is now turned upside down in the stirrup so that the scale appears inverted. It is here that the great convenience of an octagonal magnet becomes apparent, as it is possible at once to place the magnet in the stirrup in either the erect or inverted position, whereas with a round magnet in the older forms of instruments several trials are usually necessary.[a] Without changing the reading of the horizontal circle take several more readings of the scale of the magnet at intervals of one minute. Then return the magnet to the erect position and make several more scale readings. Read the horizontal circle to see that no change has taken place, remove the magnet, and conclude the set of observations by pointing on the reference mark. In general it will be found that the erect and inverted scale readings differ by several scale divisions owing to the noncoincidence of the magnetic and geometric axes of the magnet. The mean of the two gives the division of the scale corresponding to the magnetic axis, and we can then reduce the reading of the horizontal circle when pointing on the recorded scale division to what it would have been had we pointed parallel to the magnetic axis. Increasing scale readings, "magnet erect," correspond to decreasing circle readings.

The value in arc of one division of the scale is easily found by pointing on successive 5 or 10 division marks and noting the corresponding readings of the horizontal circle. In this particular instrument one division equals 2'.

The following example shows the form of record and computation. The azimuth of the mark and the reduction to local mean were obtained from the azimuth observations reproduced on pages 90 to 93. The diurnal variation or correction to reduce to mean of day was obtained from results of continuous observations at the magnetic observatory at Baldwin, Kans. In the absence of such results, an approximate correction would be obtained from a table similar to that given on page 47 (Table III), but in either case allowing for the fact that the diurnal variation increases as we go toward the magnetic pole.

---

[a] In some instruments of foreign make, recently imported by the Survey, arrangements are made whereby the round magnet can be inverted readily 180° from the outside without being obliged to open the magnetometer box and to take hold of the magnet.

TRUE MERIDIAN AND MAGNETIC DECLINATION.

Magnetic observations.   *Declination.*

Station, Paducah, Ky.
Instrument, Mag'r No. 19.
Mark, Church spire.
Magnet, 19 L.

Date, Tuesday, July 2, 1901.
Observer, W. W.

Line of detorsion, 310°.

Chron. time	Scale	Scale readings			Horizontal circle readings			
		Left	Right	Mean			Mark	Magnet
h m		d	d	d			° ′	° ′
7 54	E	38.1	38.8	38.45	Before	A	328 00.0	178 45.5
55	E	37.9	39.1	38.50		B	147 56.5	358 44.5
57	I	37.7	37.0	37.35	After	A	328 00.0	178 45.5
58	I	37.7	37.0	37.35		B	147 56.5	358 44.5
59	I	37.7	37.0	37.35	Mean		327 58.2	178 45.0
8 00	I	37.7	37.0	37.35				
02	E	38.1	38.6	38.35	Scale erect, mean			d 38.40
03	E	38.0	38.6	38.30	Scale inverted, mean			37.35
					Axis			37.88

Mean scale reading erect		38.40
Axis		37.88
Scale—axis		+0.52
Reduction to axis		+1′.0
Circle reading		178 45.0
Mag'c S. M. reading		178 46.0
Mark reading		327 58.2
Azimuth of mark [a]		153 40.4
True S. M. reading		174 17.8
Magnetic declination E.		4 28.2
Diurnal variation		−2.9
Mean declination E.		4 25.3

Remarks:
   Bright, clear day
   Temp. 33°.5 Cent.
   Torsion weight suspended 20 minutes

	h m
Mean chron. time	7 58.5
Chron. fast on L. M. T.	54.5
Local mean time	7 04

[a] Counted from South around by West from 0° to 360°.

CPSIA information can be obtained
at www.ICGtesting.com
Printed in the USA
LVOW13*2350251116
514292LV00020BB/437/P